U0058299

一定可以輕鬆學會

三明治・鬆餅・醬料

100種

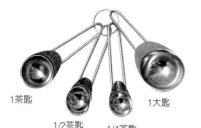

1茶匙
1/2茶匙　1/4茶匙　1大匙

量杯/量匙換算

1公升＝1000毫升
1毫升＝1cc
1杯＝240cc＝16大匙
1大匙＝3茶匙＝15cc
1茶匙＝1小匙＝5cc
1/2茶匙＝2.5cc
1/4茶匙＝1.25cc

公克/台斤換算

1公斤＝1000公克＝0.6台斤
1台斤＝16兩＝600公克
1兩＝37.5公克
1磅＝454公克＝12兩

不同材料的量杯量匙與重量對照表

名　稱	量　　杯	1 大 匙	1 茶 匙	1/2 茶 匙	1/4 茶 匙
鹽	180公克	14公克	4~5公克	2~3公克	1~1.5公克
砂糖	180公克	15公克	4公克	2公克	1公克
細砂糖	108公克	9公克	3公克	1.5公克	0.75公克
醋	250公克	15公克	5公克	2.5公克	1.25公克
油	228公克	15公克	5公克	2.5公克	1.25公克
太白粉	120公克	7公克	2.5公克	1.25公克	0.6公克
麵粉	120公克	7公克	2.5公克	1.25公克	0.6公克

麵包＋抹醬＋內餡＝無限可能的三明治！

在工作忙碌中，有時候連用餐時間都沒有，或是大魚大肉吃膩了，想換換胃口，快又好吃且又能填飽肚子的三明治就是最好的選擇！

製做三明治的時候必須要注意到幾個步驟及簡易器具。首先要想到的是主要材料(Body)，如肉類、海鮮或是乳酪、蔬菜…等。再來是裝飾配料(Graish)，如泡菜類、酸豆或是萵苣類生菜…等。其次是選擇麵包(Base)，如吐司、法國麵包或是自製鬆餅、薄餅…等。最後搭配抹醬(Spread)，如美奶滋、芥茉醬、蕃茄醬或是果醬、花生醬都可以，可依自己喜愛的材料及麵包來搭配，且是創意十足，變化萬千的三明治。

在這本「三明治、鬆餅＆醬料100道 Creative Easy Sandwich」書中，三明治分成冷式三明治(Cold Sandwiches)、熱三明治

(Hot Sandwiches)、填入及包捲三明治(Stuffed and Rolled Sandwiches)和開面式三明治(Open Face Sandwiches)，還有加上自製鬆餅、煎餅、薄餅及可麗餅等多種樣式的三明治。不但外層的麵包造型多樣，內層的餡料組合更是超過你想像的豐富，每種均有中英對照，詳細的麵包、材料、配料、抹醬、作法…，更有煎餅、鬆餅、薄餅、可麗餅的麵糊完整配方、製作圖解，讓您一目瞭然零失敗。

書中還有麵包、吐司與內餡、醬料的搭配：三明治製作的基本材料器具…等完整介紹，絕對是一本可讓廚房新手輕鬆入門的三明治範本，對於料理好手而言，更期望它能成為讓您從中找到許多獨具創意及中西變化的三明治寶典。

陳寬定Eddie Chen
國立高雄餐旅學院
西餐廚藝系

經歷Work Experience

1982～1989	Hilton international Taipei Saucier 台北希爾頓大飯店師傅
1989～1996	Grand Hyatt Taipei Executive Souse Chef 凱悅大飯店行政副總廚
1996～	Kai ping high school 開平高中教師
1997～ Now	Kaohsiung Hospitality College, Assistant professor 國立高雄餐旅學院助理教授
1988～1992	救總職訓所教師
1991	Council of Labor Affairs, Executive Yuan Taipei R. O. C. Vocational Training Competition judges 行政院勞委會國際技能競賽裁判
1995	33rd international Vocational Training Competition judges Lyon France 法國里昂國際技能競賽裁判
1995	Jin won institute of technology 景文技術學院兼任教師
1996	Shin Chien University 實踐大學生活應用科學系兼任教師
1997	Attended a WACS judges Bangkok 泰國曼谷國際美食大賽裁判
2002	Council of Labof Affairs. Executive Yuan Taiwan R. O. C. International vocational Training Competition judges. 行政院勞委會國際技能競賽裁判長
2005	World Skills Competition Helsinki, Finland Judges. 芬蘭赫爾辛基國際技能競賽裁判

獲獎Awards

1986	Golden chef National cooking competition Taipei Golden medal 台北金廚獎金牌
1986	International Food competition Singapore, Team Golden medal 新加坡美食展隊金牌牌
1987	Golden Chef National cooking competition Taipei Golden medal 台北金廚獎金牌
1988	International Kochkunst-Auss TELLung, Two Golden medal 德國法蘭克福奧林匹克雙金牌
1989	Taipei Chinese Food Festival Golden medal 中華美食展金牌
1994	第二十八屆觀光旅館業優良廚師從人員
1995	第二十九屆觀光旅館業模範廚師
1996	Taipei Chinese Food Festival Student Team Golden medal 中華美食展校園傳承金牌
1997	Taipei Chinese Food Festival Student Team Golden medal 中華美食展校園傳承金牌
2001	The 8th Taiwan R. O. C. Technical model Selection Committee. Top 10 Techincal medel. 中華民國第八屆十大技術楷模金技獎
2004	FHC 6th China International Culinary Arts Competition The Team Golden Medal 上海美食展團隊金牌獎
2004	Taipei Chinese Food Festival South Team good medal 中華美展南區隊金牌

國際合作Internship

1989	Hilton International Hong Kong
1992	Hyatt Regency Adeaide Australia
1993	Hyatt Regency Singapore
1998	Culinary Institute of America
2004	Culinary Institute of America

感謝以下學生們的協助！

吳佩霜	黃鈺錚
廖凡琇	曾詩韻
何文毅	徐佑禮
薛名宏	蘇育弘
吳奎璋	宋介源
馮鈺雅	朱振江
劉韋琳	施哲文
張佩雯	簡儀松
曾詠晴	汪韋廷
洪嘉君	許雅晴
徐嬋秀	

蕃茄醬 Ketchup

可直接買到市售罐裝品，以蕃茄為主要原料，其餘的調味可能因不同的製造商，而有不太相同的配方，基本原料是番茄、醋、糖、鹽、丁香，肉桂、洋蔥、芹菜和其它蔬菜也常常加入。

日式芥茉醬 Wasabi

是由山葵塊根磨成的粉末製成，一般可直接買到市售膏狀的軟管芥茉醬或是粉裝的芥茉粉，再用水調勻而成。書中的日式芥茉醬是用市售的芥茉醬再加入少許的美奶滋而成。

義式甜醋汁 Balsamic Dressing

義大利甜醋 30毫升、橄欖油 90毫升、洋蔥碎 30公克、大蒜碎 10公克、羅勒碎 5公克、鹽和胡椒適量
作法：將所有材料混合均勻即可。

白蘭地醬 Brandy Sauce

美奶滋 150毫升、蕃茄醬 45毫升、洋蔥碎 50公克、酸黃瓜碎 30公克、蛋碎 1顆、巴西里碎 5公克、白蘭地 25毫升
作法：將所有材料混合均勻即可。

狄蓉芥茉醬 Dijon Mustard

產自法國狄蓉地區的狄戎芥茉醬是用棕芥茉子磨碎，和水、白酒、鹽、香料 混合製作出的，味道卻是濃烈辛辣，可直接買到市售罐裝品，在三明治中適合與肉類搭配，可以與法式芥茉醬替換使用。

法式芥茉醬 French Mustard

可直接買到市售罐裝品，多與美式的三明治中搭配，像是熱狗堡、漢堡…等，可以與狄蓉芥茉醬替換使用。

梅醬 Plum Sauce

本書中使用的是中式的梅醬，很適合與烤鴨搭配製作三明治。超市可直接買到以醃過的梅子肉加上調味料製成的罐裝品，使用非常方便。

美乃滋 Mayonnaise

可直接買到市售罐裝品，是由蔬菜油、蛋黃、檸檬汁或醋，還有其他調味料所製成濃稠乳脂狀的調味醬。

沙沙醬 Salsa

蕃茄丁250公克、檸檬汁15毫升、洋蔥碎50公克、大蒜碎10公克、
辣椒碎10公克、香菜10公克、檸檬皮碎1/2顆、
橄欖油10毫升、鹽和胡椒適量
作法：將所有材料混合均勻即可。

花生醬 Peanut butter

花生醬不含膽固醇，兩大匙含10克蛋白質、維他命B(包括葉
酸)，鉀，鎂以及纖維及190kcal熱量和16克脂肪，不僅很適合
製作三明治，適量的攝取更是素食者絕佳的平價蛋白質來源。

蜜洋蔥 Onion Comfit

洋蔥絲 300公克、紅酒醋 25毫升、紅酒 150毫升、糖 50公克
作法：1.以少許的奶油將洋蔥絲炒至軟且呈現焦糖色。
　　　2.加入紅酒並使其酒精蒸散後，放入糖和紅酒醋，稍煮一下即可。

酪梨醬 Guacamole

酪梨2顆、檸檬汁15毫升、洋蔥碎50公克、大蒜碎10公克、
辣椒碎10公克、香菜碎10公克、檸檬皮碎1/2顆、鹽和胡椒適量
作法：1.酪梨去皮去籽後切成塊狀,放入鋼盆中壓碎。
　　　2.混合其他材料即可。

濃縮義大利甜醋醬 Balsamic Reduced

義大利甜醋 100毫升

作法：視需要的量,將兩倍的醋濃縮至一倍即可。

奶油起士 Cream Cheese

奶油起士 250g、大蒜碎 10g
作法：將奶油起士放進鋼盆中,以打蛋器稍壓使其軟化。
　　　加入大蒜碎,攪拌均勻即可。

酸奶油以新鮮牛奶加入酵母菌發酵製成，口感滑順綿密，入口即化
且清爽不油膩，酸味與奶香兼具，常應用在各式歐美料理中，與醃
燻鮭魚或其他重口味食材搭配，超市中可買到現成罐裝的酸奶油。

打發鮮奶油 Whipped Cream

直接將濃稠液體狀的鮮奶油Cream以網狀攪拌器打發(打
入空氣)即可，打發的鮮奶油會呈現細緻的花紋，多用在
甜口味的鬆餅、薄餅…等料理中。

高麗菜沙拉 Cole slaw

材料：

高麗菜絲.............................300公克
鹽.......................................3公克
美奶滋.................................80毫升
鹽和胡椒.............................適量

做法：

將高麗菜絲與鹽混合使其軟化,待出水後,將水擠乾加入美奶滋,以鹽和白胡椒調味即可。

Ingredients
Cabbage julienne 300gr
Salt 3gr
Mayonnaise 80ml
S&P to taste

Methods ：
Combine salt and cabbage julienne until julienne become soft. Squeeze
out the water. Mix well with mayonnaise、S&P. Serve as need.

鮪魚沙拉
Tuna fish salad

材料：

鮪魚罐 1罐
洋蔥碎.................................50公克
西芹碎.................................30公克
酸黃瓜碎.............................30公克
巴西里碎.............................5公克
鹽、白胡椒 ....適量

做法：

① 將鮪魚罐頭去油備用。
② 與剩餘材料切碎與鮪魚混合,再放入美奶滋拌勻並調味即可。

Ingredients
Tuna fish canned 1TIN
Onion chopped 50gr
Celery chopped 30gr
Pickle chopped 30gr
Parsley chopped 5gr
S&P as Need

Methods ：
1. Take out the tuna fish oil from the can.
2. Combine all left items with tuna fish in the mixing
   bowl, mix well. Add mayonnaise, then toss and mix
   well, season to taste.

冷 式 三 明 治 Cold Sandwich

潛水艇三明治 01
Submarine

主要材料：
火腿片...........................2片
義式香腸片...................5片
起士片...........................3片

裝飾材料：
美生菜.....................60公克
洋蔥絲.....................50公克
蕃茄片...........................1顆
酸黃瓜片.................30公克

麵包：
法國麵包.....................半條

抹醬：
白蘭地醬..................50毫升
狄蓉芥茉醬..............20毫升
蕃茄醬.....................20毫升

Ingredients
Body：
Ham sliced   2p.c
Salami sliced 5p.c
Gruyere cheese 3p.c

Garnish：
Iceberg lettuce 60gr
Onion sliced 50gr
Tomato sliced 1EA
Pickle sliced 30gr

Bread：
French bread 1/2 loaf

Spread：
Brandy Sauce 50ml
Dijon Mustard 20ml
Ketchup 20ml

做法：
① 將法國麵包切開先抹上白蘭地醬，再放入裝飾材料。
② 將主要材料放上排好看。
③ 食用時，可淋上芥茉醬及蕃茄醬。

Methods：
1. Cut the French bread, spread brandy sauce on cutting surface of bread, and then put on all garnish items.
2. Add all body items on top.
3. Sprinkle some mustard, ketchup for service.

義式香腸 Salami
主要以豬肉、牛肉、蒜頭和其他香料製作的義式香腸，極少煙燻，通常醃漬後便風乾處理，其形狀、大小和種類繁多。

煙燻火雞肉片 Smoked Turkey Ham
將火雞肉醃漬(有些產品會經過醃燻的過程)後，真空保裝出售。有時可買到切片的包裝。

宴會式三明治
Hero Sandwich Party Style

主要材料：
火雞肉片 5片
牛肉火腿片 5片
火腿片 5片
水煮蛋切片 2顆

裝飾材料：
美生菜 80公克
紫包心菜 50公克
蕃茄片 2顆
小黃瓜片 1顆
酸黃瓜碎 1顆
黑橄欖碎 10顆

麵包：
長條法國麵包 1條

抹醬：
美奶滋 200毫升
狄蓉芥茉醬 50毫升

做法：
① 將主要材料切片備用，將長條法國麵包切開，抹上抹醬。
② 放入裝飾材料，再加入主要材料排好即可享用。

Ingredients
Body：
Turkey Ham sliced 5p.c
Pastrami sliced 5p.c
Ham sliced 5p.c
Hard Boiled Egg sliced 2EA

Garnish：
Iceberg lettuce 80gr
Reddish lettuce 50gr
Tomato sliced 2EA
Baby cucumber 1EA
Pickle chopped 1EA
Black olive chopped 10EA

Bread：
French Bread Big loaf
1 loaf

Spread：
Mayonnaise 200ml
Dijon mustard 50ml

Methods ：
1. Cut the French bread, spread brandy sauce on cutting surface of bread, and then put on all garnish items.
2. Add all body items on top
3. Can be carving service.

法國麵包 French Bread
一種以麵粉、水、酵母和鹽烘烤成的脆皮麵包，理想的法國麵包為外脆內軟，且表面有不規則的裂痕。通常搭配果醬、奶油或是當佐餐麵包享用。如果吃不完可用保鮮膜完全包裹後放入冷凍庫，可保存1～2個月，食用前用烤箱加熱即可。

03 鮪魚沙拉三明治 Tuna fish salad Sandwich

主要材料：

鮪魚罐......................1罐

裝飾材料：

洋蔥碎 50公克
西芹碎 30公克
酸黃瓜碎............30公克
巴西里碎...............5公克
鹽、白胡椒............適量

麵包：

白吐司......................3片

抹醬：

美奶滋 150毫升

做法：

① 將主要材料去油備用。
　 將裝飾材料切碎加入
　 主要材料中，再放入
② 美奶滋拌勻即可。
● 將拌好的材料塗抹麵
　 包上即可。

Ingredients
Body：
Tuna fish canned 1TIN

Garnish：
Onion chopped 50gr
Celery chopped 30gr
Pickle chopped 30gr
Parsley chopped 5gr
S&P as Need

Bread：
White Toast 3p.c

Spread：
Mayonnaise 150ml

Methods ：
1. Take out the tuna fish oil from the can.
2. Add all the garnish items and tuna fish in the mixing bowl, mix well. Add mayonnaise, then toss and mix well.
3. Spread the mixture of all ingredients on the white toast. Serve as need.

04 火腿起士三明治 Ham and cheese Sandwich

主要材料：

火腿片......................2片
起士片......................2片

裝飾材料：

黑橄欖......................3顆

麵包：

全麥吐司..................4片

抹醬：

美奶滋 60毫升
狄蓉芥茉醬.........20毫升

做法：

① 將主要材料切片備用。
② 將全麥吐司（可烤上
　 色或不烤）抹上美奶
　 滋芥茉醬，再上主要
　 材料即可。

Ingredients
Body：
Ham sliced 2p.c
Gruyere cheese 2p.c

Garnish：
Black Olive 3EA

Bread：
Brown Toast 4p.c

Spread：
Mayonnaise 60ml
Dijon Mustard 20ml

Methods ：
1. Cut all the body items into slices.
2. Spread the mayonnaise and mustard on the brown toast (Toasted or not) than put on all the body items.

總匯三明治 Club Sandwich with French Fries

主要材料：
水煮雞胸肉切片..70公克
烤培根......................2片
水煮蛋切片..............1顆

裝飾材料：
美生菜...............60公克
蕃茄切片.................1顆

麵包：
白吐司....................3片

抹醬：
美奶滋.............180毫升

做法：
① 將雞胸肉煮熟切片，培根烤好（可依個人喜好烤脆與否）蛋煮熟去殼切片。
② 將白吐司烤上色，塗抹上美乃滋，放上裝飾材料再放上主要材料。
③ 四邊用牙簽固定切成四片，放上炸薯條即可。

Ingredients
Body：
Chicken Breast
poached 70gr
Bacon baked 2p.c
Hard Boiled Egg sliced
1EA

Garnish：
Iceberg lettuce 60gr
Tomato sliced 1EA

Bread：
White Toast 3p.c

Spread：
Mayonnaise 180ml

Methods：
1. Poached chicken breast, than cut to sliced .
   Baked the bacon (Optional, crispy or not) and
   boiled the egg than cut to sliced.
2. Toasted the white toast and spread mayonnaise, put on all garnish and body items.
3. Cut into 4 parts, served with French fries.

06 藍乳酪蘋果沙沙三明治
Blue cheese and Apple Salsa with Walnut Sandwich

主要材料：

藍乳酪起士.........60公克
蘋果沙沙醬.......120公克
┌ 蘋果切小丁..........1顆
│ 檸檬汁.............5毫升
│ 洋蔥碎...........10公克
│ 香菜碎.............3公克
│ 大蒜碎.............3公克
└ 鹽、白胡椒.......適量

裝飾材料：

核桃碎................50公克

麵包：

全麥麵包.................2片

抹醬：

軟化奶油............30公克

Ingredients
Body：
Blue cheese 60gr
Apple salsa 120gr
┌ Apple small diced
│ 1EA
│ Lemon juice 5ml
│ Onion chopped 10gr
│ Coriander chopped
│ 3gr
│ Garlic chopped 3gr
└ S&P as Need

Garnish：
Walnut chopped 50gr

Bread：
Rye Bread 2p.c

Spread：
Melted Butter 30gr

Methods：
1. Slice the blue cheese, mix up all the ingredients of apple salsa.
2. Toast walnuts and chop.
3. Spread melted butter on rye bread, put on all the body and garnish ingredients. Serve as need.

做法：

① 將藍乳酪起士切片，蘋果切丁調成沙沙醬。

② 將核桃烤熟切碎。

③ 將全麥麵包塗抹奶油，再放上主要材料，裝飾材料即可。

07 英式蛋三明治
Ham and cheese Sandwich

主要材料：

水煮蛋切碎..............5顆
青蔥碎 25公克

麵包：

白吐司......................2片

抹醬：

美奶滋 120毫升

Ingredients
Body：
Hard Boiled Egg
chopped 5EA
Spring onion chopped
25gr

Bread：
White Toast 2p.c

Spread：
Mayonnaise 120ml

做法：

① 將蛋煮熟去殼切碎加入青蔥碎及美乃滋拌勻，再塗抹於白吐司上即可。

② 可以個人切成自己喜好的形狀。

Methods：
1. Boiled the egg to well down then Spring onion chopped and mix with mayonnaise. Spread on the white toast.
2. Cut freely by following your own style.

華爾道夫沙拉三明治
Waldorf salad Sandwich

主要材料：

蘋果 1顆
西芹丁 60公克
葡萄乾 20公克
煙燻火雞肉片 50公克

裝飾材料：

核桃 20公克
西芹葉 5公克
鹽、白胡椒 ....適量

麵包：

軟式麵包 1個

抹醬：

美奶滋 80毫升

做法：

① 將蘋果去皮切丁，西芹切丁，加入葡萄乾，一半的核桃切碎，拌入美奶滋及鹽、胡椒調味。

② 放入軟式麵包灑上另一半的核桃，及西芹葉裝飾即可。

Ingredients
Body：
Apple 1EA
Celery diced 60gr
Raisins 20gr
Smoked Turkey Ham 50gr

Garnish：
Walnut 20gr
Celery leaves 5gr
S&P TO TASTE
Bread：
Soft Bread 1EA

Spread：
Mayonnaise 80ml

Methods：

1. Peel and core the apple, cut into dices. Combine apple dices、celery dices、raisins and chopped walnuts, then add mayonnaise、salt、pepper and mix well.

2. Put Waldorf salad and Smoked Turkey Ham on the soft bread, sprinkle with some walnuts chopped. Garnish withcelery leaves. Serve as need.

09

煙燻鮭魚三明治
Smoked Salmon Sandwich

主要材料：
煙燻鮭魚 5片

麵包：
馬芬麵包 2顆

裝飾材料：
洋蔥絲 10公克
蕃茄片 1顆

抹醬：
白蘭地醬 50毫升

Ingredients
Body：
Smoked Salmon 5p.c

Garnish：
Onion sliced 10gr
Tomato sliced 1EA

Bread：
Muffins Bread 2EA

Spread：
Brandy Sauce 50ml

做法：
① 將煙燻鮭魚切片，蕃茄切片，洋蔥切絲備用。
② 馬芬麵包抹上白蘭地醬，再將主要及裝飾材料放上即可。

Methods：
1. Slice smoked salmon、tomato and onion.
2. Spread brandy sauce on muffins, put on body and garnish items. Serve as need.

馬芬麵包Muffins
一種以杯狀烤盤烤出的快速柔軟麵包，其麵糊通常以堅果或是水果提味。

煙燻火雞肉三明治 10

Smoked Turkey Sandwich

主要材料：
煙燻火雞肉片 5片

麵包：
軟式麵包 2顆

裝飾材料：
蜜洋蔥 50公克
羅莎生菜 1片

抹醬：
軟化奶油......30公克

做法：
將軟麵包抹上奶油放上生菜，再放上火雞肉及蜜洋蔥即可。

Ingredients
Body：
Smoked Turkey Ham
5p.c

Garnish：
Onion comfit 50gr
Lollo rosso 1p.c

Bread：
Soft Rolls 2EA

Spread：
Melted Butter 30gr

Methods：
Spread butter on soft rolls. Lay lettuce and put turkey ham and onion comfit on top.

漢堡包 Hamburger
直徑約為9～10公分的圓形麵包，
質地鬆軟且蓬鬆，可以一般麵粉或
全麥麵粉來製作，麵包頂端通常
以芝麻粒、罌粟籽來裝飾。

11 燻雞胸三明治
Smoked Chicken Breast Sandwich

主要材料：
燻雞胸切片............120公克

裝飾材料：
綜合脆綠沙拉 60公克
義式甜醋汁..............20毫升

麵包：
法式麵包......................1個

抹醬：
軟化奶油 50公克

做法：
① 將燻雞肉切片備用。
② 將綜合脆綠沙拉抹上義式甜醋汁。
③ 將法國麵包抹上奶油，再放上主要材料及裝飾材料即可。

蘿蔓生菜 Romaine Lettuce
形狀長且頭細，葉片鬆散，口感酥脆，又名萵苣菜。枝葉末端顏色較暗呈現暗綠色葉片，中心較顏色較淡呈淡黃色，味道微苦，葉柄易碎。

Ingredients
Body：
Smoked Chicken Breast
sliced 120gr

Garnish：
Mixed Green Salad
60gr
Balsamic Dressing
20ml

Bread：
French bread 1EA

Spread：
Melted Butter 50gr

Methods：
1. Slice the smoked chicken breast.
2. Combine green salad and balsamic dressing. Mix well and toss.
3. Spread butter on French bread. Put on salad and smoked chicken breast slices. Serve as need.

12

草莓乳酪三明治
Cream cheese and strawberries Sandwich

主要材料：
草莓 80公克

裝飾材料：
洋蔥碎 10公克
大蒜碎 3公克
香菜碎 3公克
鹽、白胡椒............適量

麵包：
馬芬麵包 2顆

抹醬：
奶油起士.............80公克

做法：
① 將草莓切丁加入裝飾材料拌勻。
② 將馬芬麵包抹上奶油起士，再放上主要材料，及裝飾材料即可。

Ingredients
Body：
Strawberries 80gr

Garnish：
Onion chopped 10gr
Garlic chopped 3gr
Coriander chopped 3gr
S&P TO TASTE

Bread：
Muffins Bread 2EA

Spread：
Cream Cheese 80gr

Methods：
1. Cut strawberries into dices, and combine with garnish items.
2. Spread cream cheese on muffin, put on the mixture of strawberries and garnish. Serve as need.

Peanut Butter Bagel 花生醬培果 **13**

主要材料：
花生醬 80毫升

麵包：
培果 1顆

抹醬：
奶油起士............80公克

做法：
將培果切開抹上花生醬及奶油起士即可。

Ingredients
Body：
Peanut Buttes 80ml

Bread：
Bagel 1EA

Spread：
Cream Cheese 80gr

Methods：
Cut the bagel, and spread peanut butter and cream cheese on the cutting surface. Serve as need.

14

燻鰻魚三明治 Smoked Trout Sandwich

主要材料：
燻鰻魚 1/2條
（約80公克）

裝飾材料：
蜜洋蔥 50公克
洋蔥絲 50公克

麵包：
軟式麵包 2顆

抹醬：
美乃滋 80毫升

Ingredients
Body：
Smoked Eel 1/2EA
（80gr）

Garnish：
Onion comfit 50gr
Onion sliced 5gr

Bread：
Softed Bread 2EA

Spread：
Mayonnaise 80ml

做法：
① 將燻鰻魚去皮去骨備用。
② 將麵包抹上美奶滋，再放上蜜洋蔥，及燻鱒魚即可。

Methods：
1. Take off the bone and skin of Eel.
2. Spread mayonnaise on soft bread, then put on onion comfit and smoked trout. Serve as need.

紫紅生菜radicchio
又稱為義大利菊苣，有著紫
紅的葉片，葉片根部為白色
的帽狀菜片，稍帶有苦味，
適合用於沙拉或當盤飾。

義式香腸三明治

Salami and Mozzarella cheese Sandwich

主要材料：
義式香腸切片....................10片

裝飾材料：
莫札瑞拉起士 80公克
蕃茄片................................1顆

麵包：
法式麵包............................1顆

抹醬：
美奶滋.........................50毫升
狄蓉芥茉醬...................15毫升

Ingredients
Body：
Salami sliced 10p.c

Garnish：
Mozzarella cheese 80gr
Tomato sliced 1EA

Bread：
French bread 1EA

Spread：
Mayonnaise 50ml
Dijon Mustard 15ml

做法：
① 將法式麵包切開抹上美奶滋芥茉醬。
② 再將主要材料及裝飾材料放上即可。

Methods：
1. Cut French bread, then spread mayonnaise and mustard on the cutting surface.
2. Put on body and garnish items. Serve as need.

16 鮮蝦三明治
Shrimp and Cole slaw Sandwich

主要材料：
草蝦...................6顆

裝飾材料：
高麗菜沙拉..80公克
（→p.10）
美生菜...............1片

麵包：
軟式麵包 1顆

抹醬：
美奶滋 80毫升

做法：
將軟式麵包切開放上主
要及裝飾材料即可。

Ingredients
Body：
Shrimp 6EA

Garnish：
Cole slaw 80gr （→p.10）
Green Lettuce 1p.c

Bread：
Soft Bread 1EA

Spread：
Mayonnaise 80ml

Methods：
Cut the soft bread. And put body and garnish items on top, serve as need.

脆綠沙拉培果三明治

17

Green Salad in Bagel

主要材料：
脆綠沙拉....120公克
（各種綜合生菜）

麵包：
培果.................1個

裝飾材料：
義式甜醋汁 ..40毫升

抹醬：
奶油起士......80公克

做法：
將培果烤上色，抹上奶油起士，再放上主要材料拌裝飾材料即可。

Ingredients
Body：
Green salad 120gr

Garnish：
Balsamic Dressing
40ml

Bread：
Bagel 1EA

Spread：
Cream Cheese 80gr

Methods：
Toast the bagel until golden brown. Spread cream cheese on top. Put on body and garnish items on top. Serve as need.

捲鬚菜Frisée
葉片呈淡黃綠色，味道
非常苦。

27

18 下午茶三明治
Tea Time Sandwich

主要材料：

起士片.....................4片
火腿片.....................4片
鮪魚沙拉...............200g
（→p.10）
蛋沙拉 200g
（→p.18）

裝飾材料：

小黃瓜片 2片
海苔 1片

麵包：

白吐司.....................4片
全麥吐司................4片

做法：

① 將白吐司分別放上蛋沙拉和鮪魚沙拉，去邊後可切成長方形或是三角形等喜好的形狀，分別以小黃瓜片和海苔片裝飾後即可。

② 在全麥吐司依序放上起士片、火腿片、起士片，去邊後可切成長方形或是三角形等喜好的形狀即可。

Ingredients
Body：
Cheese slices 4p.c.
Ham slices 4p.c.
Tuna salad 200g
(→p.10)
Egg salad 200g
(→p.18)

Garnish：
Baby cucumber slices 2p.c.
Nori 1 p.c.

Bread：
White toast 4 p.c.
Brown toast 4 p.c.

Methods：
1. Put egg salad or tuna salad on white toast, cut off the crust and cut into the shapes which you preferred. Garnish with baby cucumber slice and nori.
2. Put cheese slices and ham slices on brown toast, cut off the crust and cut into the shapes which you preferred. Serve as need.

鮪魚馬芬三明治 19
Tuna fish Salad in muffins

主要材料：
鮪魚沙拉 ..150公克(→p.10)

裝飾材料：
水煮蛋...........................1顆
蕃茄片...........................3片

麵包：
馬芬麵包.......................1顆

抹醬：
美奶滋....................150毫升

做法：
將鮪魚沙拉及裝飾夾入馬芬
麵包即可。

Ingredients
Body：
Tuna fish salad 150gr
(→p.10)

Garnish：
Hard Boiled 1gg
Slice 1EA
Tomato Slice 3p.c

Bread：
Muffins Bread 1EA

Spread：
Mayonnaise 150ml

Methods：
Put the Tuna fish salad and
garnish items on muffin.
Serve as need.

粗鹽 Kosher Salt
未經過精緻的海鹽，呈現不規則
狀的結晶體，鹹味不像精鹽那般
強烈，稍帶點海水的淡淡腥香。

29

20

綜合水果
培果三明治
Mixed Fruit in Bagel

主要材料：
草莓.................3顆
蘋果.............1/2顆
鳳梨...........50公克
哈蜜瓜........50公克

裝飾材料：
柳橙汁.........10毫升
蜂蜜............50毫升

麵包：
培果.................1個

抹醬：
柳橙果醬......80公克

做法：
① 將主要材料與裝飾材料拌好。
② 再將培果切開抹上柳橙果醬，再放上主要材料即可。

Ingredients
Body：
Strawberry 3EA
Apple sliced 1/2EA
Pineapple 50gr
Honey melon 50gr

Garnish：
Orange juice 10ml
Honey 50ml

Bread：
Bagel 1EA

Spread：
Orange jam 80gr

Methods：
1. Combine body and garnish items.
2. Cut the bagel, and spread orange jam on cutting surface. Put on the mixture of fruits. Serve as need.

培果 Bagel
為猶太教的傳統麵包，為甜甜圈狀，質地相當緊密，大都是先置水中烹煮後再放到烤箱烘烤，外皮亮麗，耐咀嚼。外表可沾裹芝麻、罌粟籽等裝飾物。

可頌三明治 21
Croissant Sandwich

主要材料：
鮪魚沙拉 500g
（→p.10）
蕃茄片 5片

麵包：
可頌麵包..............5個

抹醬：
美奶滋...............50ml

做法：
① 將可頌麵包剖開,在切面處抹上美奶滋。
② 在麵包內先墊上一片番茄片,放上鮪魚沙拉即可。

Ingredients
Body：
Tuna Salad 500g
（→p.10）
Tomato sliced 5ea

Bread：
Croissant 5 EA

Spread：
Mayonnaise 50ml

Methods：
1. Cut open croissants, spread mayonnaise on cutting surface.
2. Put tomato slice on croissants, and put some tuna salad on tomato slices. Serve as need.

可頌麵包 Croissant
形狀大多呈牛角狀,味道相當濃郁且具奶油味,為一種多層次麵包。

熱三明治

Hot Sandwich

熱狗堡
Hot dog

主要材料：
熱狗腸......................1顆

裝飾材料：
洋蔥碎 50公克
酸黃瓜碎............20公克

麵包：
軟麵包......................1顆

抹醬：
白蘭地抹醬.........30毫升
法式芥茉醬.........20毫升
蕃茄醬 20毫升

做法：
① 將熱狗腸烤好，或煎好備用。
② 將軟麵包切開抹上白蘭地抹醬，再放上裝飾材料，及烤好或煎好的熱狗腸。
③ 淋上芥茉醬及蕃茄醬即可。

Ingredients
Body：
Hot dog Sausage 1EA

Garnish：
Onion chopped 50gr
Pickle chopped 20gr

Bread：
Soft Bread 1EA

Spread：
Brandy Sauce
French Mustard
Ketchup

Methods：
1. Roast hot dog sausage in oven. (Or pan fry)
2. Cut open the soft bread and spread brandy sauce on cutting surface, put on garnish items and cooked sausage.
3. Sprinkle some mustard and ketchup on top. Serve as need.

23

碳烤牛肉三明治 Grilled Beef Sandwich

主要材料：
沙朗牛肉....120公克
鹽、白胡椒......適量

裝飾材料：
蕃茄丁 80公克
檸檬汁 10毫升
大蒜碎 5公克
香菜碎 5公克
洋蔥碎 20公克
鹽、白胡椒......適量

麵包：
軟麵包 1顆

抹醬：
美奶滋 50毫升

Ingredients
Body：
Beef sirloin 120gr
S&P TO TASTE

Garnish：
Tomato diced 80gr
Lemon juice 10ml
Garlic chopped 5gr
Coriander chopped 5gr
Onion chopped 20gr
S&P TO TASTE

Bread：
Softed Bread 1EA

Spread：
Mayonnaise 50ml

做法：
① 將沙朗牛肉切片，灑上鹽、胡椒，烤好備用。
② 將麵包切開抹上美奶滋，放上裝飾材料，及烤好的牛肉即可。

Methods：
1. Slice sirloin, sprinkle some salt and pepper on top. Roast for use.
2. Cut open the bread and spread mayonnaise on cutting surface, put on garnish items and cooked beef. Serve as need.

焗烤火腿乳酪三明治 **24**

Gratinated Beef sliced and Mushroom Sandwich

主要材料：
牛腓肋片......120公克
洋菇片 80公克
莫札瑞拉起士 60公克

裝飾材料：
洋蔥絲 50公克
鹽、白胡椒 ....適量

麵包：
軟麵包 1顆

抹醬：
軟化奶油 30公克

做法：
① 將軟麵包切開抹上奶油烤上色備用。
② 將牛肉及洋菇片，洋蔥絲炒好放上烤好的軟
麵包，再放莫窄力拉起士，並焗上色即可。

Ingredients
Body：
Beef Tenderloin sliced
120gr
Mushroom sliced 80gr
Mozzarella cheese 60gr

Garnish：
Onion sliced 50gr
S&P TO TASTE

Bread：
Softed Bread 1EA

Spread：
Melted Butter 30gr

Methods ：

1. Cut opens the soft bread and spread butter on cutting surface. Toast in oven until colored.
2. Saute beef slices、mushroom slices and onion slices until cooked. Put cooked ingredients on top of bread. Sprinkle some mozzarella cheese on top and gratin until colored. Serve as need.

25

煎烤火腿乳酪三明治

Griddle Ham and Cheese Sandwich

主要材料：

火腿片 1片

葛利亞乳酪........2片

抹醬：

軟化奶油......30公克

麵包：

白吐司 2片

Ingredients
Body：
Ham sliced 1p.c
Gruyere Cheese 2p.c

Bread：
White Toast 2p.c

Spread：
Melted Butter 30gr

做法：

① 將白吐司抹上軟化奶油並煎上色。

② 放上一片起士片，再放上火腿，再放上另一片起士片，最後再蓋上另一片吐司。

③ 將做好的三明治放入烤箱(180℃)烤約5分鐘即可。

Methods：

1. Spread melted butter on white toast and panfry until colored.

2. Put a slice on cheese on toast, and then put a piece of ham. Put another piece of cheese on ham. Cover with a piece of white toast.

3. Put sandwich in oven. Finish cooking by using 180℃ for 5 minutes. Serve as need.

炸鱸魚排三明治 26

Fried Sea bass fillet Sandwich

主要材料：

鱸魚肉 80公克
蛋 1顆
麵包粉 30公克
麵粉 ....
白葡萄酒......15毫升
鹽、白胡椒......適量

裝飾材料：

生菜葉..............1片
高麗菜沙拉 ..50公克
（→p.10）

麵包：

漢堡包..............1顆

抹醬：

軟化奶油......15公克

做法：

① 將鱸魚醃上白葡萄酒、鹽、白胡椒粉，
   並沾上麵粉→蛋液→麵包粉後，炸成金
   黃色即可。

② 將漢堡包切開抹上奶油烤好，再放上裝
   飾材料及炸好的鱸魚排即可。

Ingredients
Body：
Sea bass fillet 80gr
Egg 1EA
Bread Crumb 30gr
Flour as Need
White wine 15ml
S&P to Taste

Garnish：
Green lettuce 1p.c
Cole Slaw 50gr
（→p.10）

Bread：
Burger Bun 1EA

Spread：
Melted Butter 15gr

Methods ：

1. Marinade sea bass fillet with white wine、salt and
   pepper. Breading with flour、egg and breadcrumb. Deep fry until the color go to golden brown.

2. Cut opens the burger bun and spread butter on
   cutting surface. Put on garnish ingredients and
   cooked sea bass fillet. Serve as need.

27

酥炸明蝦米三明治
Fried King Prawn and Rice Sandwich

主要材料：

明蝦 2尾
白葡萄酒 10毫升
鹽、白胡椒 ....適量
蛋 1顆
麵包粉 20公克
麵粉 20公克

裝飾材料：

生菜葉 1片
高麗菜沙拉50公克（→p.10）

麵包：

香米飯 200公克

抹醬：

美奶滋 20毫升

做法：

① 將明蝦去殼，加入白葡萄酒、鹽、白胡椒調味，沾上麵粉→蛋液→麵包粉後，炸成金黃色。

② 將香米飯做成方型二片，各抹上美奶滋，再放上裝飾材料及炸好的明蝦即可。

Ingredients
Body：
King prawn 2EA
White wine 10ml
S&P TO TASTE
Egg 1EA
Bread crumb 20gr
Flour 20gr

Garnish：
Green lettuce 1p.c
Cole slaw 50gr （→p.10）

Bread：
Steamed Rice 200gr

Spread：
Mayonnaise 20ml

Methods：
1. Take off the shell of prawn. Marinade with white wine、salt and pepper. Breading with flour、egg and bread-crumb. Deep fry until the color go to golden brown.
2. Make rice into rectangular shape. Spread mayonnaise on rice, put on garnish items and cooked prawns. Serve as need.

羅莎生菜 Lollo rosso
又名結球萵苣，顏色為淡紅且帶有少許綠色，葉尖端捲曲，邊緣呈紅色。味道稍苦。進口的梗較短儲存時間較久，台灣本地產的則較易腐敗且較多蟲。

28 蘑菇恩力蛋三明治
Mushroom Omelet Sandwich

主要材料：
蛋 3顆
洋菇 80公克

裝飾材料：
洋蔥丁 50公克
蕃茄丁 30公克
黑橄欖 5顆

麵包：
橄欖油麵包........1顆

抹醬：
芥茉醬 20毫升
蕃茄醬 20毫升

做法：
① 將橄欖油麵包對切成兩半並烤熱。
② 將主要材料及裝飾材料炒成蛋餅，並放上麵包切開即可。

Ingredients
Body：
Egg 3EA
Mushroom sliced 80gr

Garnish：
Onion diced 50gr
Tomato diced 30gr
Black olive 5EA

Bread：
Schiacciata Bread 1EA

Spread：
French Mustard 20ml
Ketchup 20ml

Methods：
1. Cut open the schiacciata bread and reheat for use.
2. Combine body and garnish ingredients and make into omelet. Put omelet on bread. Cut and serve.

烤鮪魚 / 蛋沙拉三明治 29
Grilled Tuna Salad and Egg Salad Sandwich

主要材料：
鮪魚沙拉....120公克
→ p.10
蛋沙拉.......120公克
→ p.10

麵包：
白吐司..............4片

做法：
① 在白吐司上放上鮪魚沙拉，蓋上另一塊吐司。
② 在白吐司上放上蛋沙拉，蓋上另一塊吐司。
③ 放入三明治爐中烤約三分鐘即可。

Ingredients
Body：
Tuna Salad120gr P.10
Egg Salad120gr P.10

Bread：
White toast 4 p.c

Methods：
1. Put tuna salad on white toast, and cover with another piece of toast.
2. Put egg salad on white toast, and cover with another piece of toast.
3. Put the whole sandwich into sandwich oven. Grill for 3 minutes. Serve as need.

30

中式烤鴨三明治

Chinese Roasted Duck Sandwich

主要材料：
烤鴨胸 1個

裝飾材料：
小酸黃瓜 10公克
香菜葉 5公克
青蔥 5公克
核桃碎 .... 適量

麵包：
銀絲捲 1個

抹醬：
梅醬 20毫升

做法：
① 將烤鴨切片備用。
② 將銀絲捲炸成金黃色切半並抹上梅醬，再放上主要及裝飾林料即可。

Ingredients
Body：
Roasted Duck Breast
1EA

Garnish：
Baby Cucumber pickle
10gr
Coriander leaves 5gr
Spring onion 5gr
Walnut Chopped as need

Bread：
Steamed Bun 1EA

Spread：
Plum Sauce 20ml

Methods：
1. Slice duck breast for use.
2. Deep fry steamed bun until golden brown. Cut open the bun and spread plum sauce on cutting surface. Put on body and garnish items. Serve as need.

碳烤雞胸三明治 31

Grilled chicken Breast Sandwich

主要材料：

雞胸 1個
大蒜碎 5公克
橄欖油 10毫升
鹽、白胡椒 ....適量

裝飾材料：

洋蔥絲 30公克
洋菇 30公克
美生菜 30公克
義大利甜醋汁 25毫升

麵包：

馬芬麵包 1顆

抹醬：

軟化奶油.............20公克

做法：

① 將雞胸加入大蒜碎、橄欖油、鹽、白胡椒拌勻，並烤熟，切片備用。

② 將裝飾材料拌勻，馬芬麵包切開並抹上奶油，再放上主要及裝飾材料即可。

Ingredients
Body：
Chicken Breast 1EA
Garlic chopped
Garlic chopped 5gr
Olive oil 10ml
S&P TO TASTE

Garnish：
Onion sliced 30gr

Mushroom sliced 30gr
Iceberg lettuce 30gr
Balsamic Dressing
15ml

Bread：
Muffin Bread 1EA

Spread：
Melted Butter 20gr

Methods：

1. Marinade chicken breast with garlic chopped、olive oil、salt and pepper. Roast until cooked. Slice for use.

2. Combine all of the garnish items. Cut open the muffin and spread butter on cutting surface. Put on all of the body and garnish items. Serve as need.

烤蔬菜三明治

Grilled Vegetable Sandwich

主要材料：

茄子	80公克
小黃瓜	80公克
洋菇	40公克
紅甜椒	20公克
黃甜椒	20公克
大蒜碎	10公克
橄欖油	10毫升
百里香	5公克
披薩草	5公克

裝飾材料：

義大利甜醋汁 20毫升

麵包：

法士達麵包 1個

抹醬：

軟化奶油 20公克

Ingredients
Body：
Eggplant 80gr
Baby cucumber 80gr
Mushroom Button 40gr
Red pepper 20gr
Yellow pepper 20gr
Garlic chopped 10gr
Olive oil 10ml
Thyme 5gr
Oregano 5gr

Garnish：
Balsamic Dressing 20ml

Bread：
Focaccia 1EA

Spread：
Melted Butter 20gr

做法：

① 將主要材料切片，加入大蒜碎，鹽、白胡椒、橄欖油及香料拌勻，碳烤，烤熟拌上義大利甜醋汁。

② 將麵包切半夾主要材料即可。

Methods：

1. Slice body items, combine garlic chopped、salt、pepper、olive oil and herbs. Mix well. Grill and mix well with balsamic dressing for use.

2. cut open the bread. Put all of the ingredients on bread. Serve as need.

法士達麵包 Focaccia
為一種經典的義大利麵包，傳統上是以馬鈴薯粉製成，藉由酵母的發酵膨脹，並以橄欖油和香草(迷迭香等)提味。

33

叉燒肉三明治

Barbecue Pork Sandwich

主要材料：
叉燒肉 120公克

麵包：
饅頭.................1顆

裝飾材料：
小黃瓜泡菜 ..30公克
洋蔥絲 20公克

抹醬：
梅醬............30毫升

做法：
① 將叉燒肉切片備用。
② 將饅頭切開抹上梅醬，並放上主要及裝飾材料即可。

Ingredients
Body：
Barbecue Pork
120gr

Garnish：
Baby cucumber
pickle 30gr
Onion sliced 20gr

Bread：
Steamed Bun 1EA

Spread：
Plum Sauce 30ml

Methods：

1. Slice barbecue pork for use.
2. Cut open steamed bun, spread plum sauce on cutting surface. Put on body and garnish items. Serve as need.

起士牛肉漢堡

34

主要材料：
牛絞肉 220公克
起士片 1片
洋蔥圈 3片

裝飾材料：
洋蔥碎 50公克
蛋白...................1顆
鹽...................3公克
黑胡椒粉........1公克
美生菜.............2片
酸黃瓜片 5片

麵包：
漢堡包 1顆

抹醬：
美奶滋 50毫升

做法：

① 將牛絞肉加入蛋白，洋蔥碎、鹽、黑胡椒
   拌勻並拍打成圓餅狀。

② 將做好的漢堡肉煎熟，放上起士片烤約
   3分鐘。

③ 將漢堡包切開抹上美奶滋，放上美生菜、
   黃瓜、烤好的漢堡肉及洋蔥圈即可。

Ingredients
Body ：
Beef minced 220gr
Sliced cheese 1p.c
Onion Rings 3p.c

Garnish ：
Onion chopped 50gr
Egg white 1EA
Salt 3gr

Black Pepper 1gr
Lettuce 2p.c
Pickle 5p.c

Bread ：
Hamburger Bun 1EA

Spread ：
Mayonnaise 50ml

Methods ：

1. Combine ground beef、egg white、onion chopped、Salt and pepper. Mix well and shape into round、flat burger steak.

2. Grill burger steak until cooked. Put cheese slices on top and roast for 3 minutes.

3. Cut open the hamburger bun and spread mayonnaise on cutting surface. Put on lettuce leaf、pickle slice and burger steak and onion rings. Serve as need.

薄片牛排三明治

Minute steak Sandwich with Green Salad

主要材料：
腓力牛排....120公克
鹽、白胡椒 ......適量
辣醬油 10毫升

裝飾材料：
洋蔥圈 50公克
蕃茄片 80公克
美生菜 50公克

麵包：
白吐司...............2片

抹醬：
美奶滋 80毫升

做法：

① 將白吐司烤上色抹上美奶滋備用。

② 將牛腓力拍成薄片，並加入鹽、胡椒煎上
色，再淋辣醬油。

③ 將一片白吐司放上牛排另一片放上裝飾材料
即可。

Ingredients
Body：
Beef Tenderloin
120gr
S&P TO TASTE
Worcestershire
Sauce 10ml

Garnish：
Onion Rings 50gr

Tomato sliced 80gr
Iceberg lettuce 50gr

Bread：
White Toast 2p.c

Spread：
Mayonnaise 80ml

Methods：

1. Toast white toast until colored. Spread
   mayonnaise on white toast.

2. Hit the beef by using meat hammer.
   Marinade with salt and pepper. Sprinkle
   some Worcestershire sauce on beef.

3. Put beef on one piece of toast. And put
   garnish items on another one. Serve as
   need.

軟殼蟹漢堡
Soft-shell crab Burger

主要材料：
軟殼蟹 1隻
白葡萄酒 10毫升
鹽、白胡椒.................適量

裝飾材料：
高麗菜沙拉30公克（→p.10）
生菜葉 1片

麵包：
漢堡包 1顆

抹醬：
美奶滋.....................30毫升

做法：
① 將軟殼蟹加入白葡萄酒，
　 鹽、白胡椒粉醃，再沾粉
　 炸熟備用。
② 將漢堡包切開，烤上色抹
　 上美奶滋，再放上主要及
　 裝飾材料即可。

Ingredients
Body：
Soft-shell crab 1EA
White wine 10ml
S&P TO TASTE

Garnish：
Sole Slaw 30gr
（→p.10）
Green lettuce 1p.c

Bread：
Burger Bun 1EA

Spread：
Mayonnaise 30ml

Methods：
1. Marinade soft-shelled crab with white wine、
   salt and pepper. Cover with flour and deep fry
   until color to golden brown.
2. Cut open the bread bun. Toast until colored.
   Spread mayonnaise on cutting surface. Put
   body and garnish items. Serve as need.

37

炸蚵沙拉三明治

Fried oyster Salad Sandwich

主要材料：

鮮蚵肉 80公克
白葡萄酒......10毫升
鹽、白胡椒......適量
蛋 1顆
麵包粉 20公克
麵粉 20公克

裝飾材料：

洋蔥絲 20公克
蕃茄片 20公克
美生菜絲......30公克
義式甜醋......10毫升
橄欖油 10毫升
鹽、白胡椒......適量

麵包：

法式麵包 1顆

Ingredients
Body：
Oyster 80gr
White wine 10ml
S&P TO TASTE
Egg 1EA
Bread crumb 20gr
Flour 20gr

Garnish：
Onion sliced 20gr
Tomato sliced 20gr
Iceberg lettuce 30gr
Balsamic Vinegar 10ml
Olive oil 10ml
S&P TO TASTE

Bread：
French bread 1EA

做法：

① 將蚵洗淨調入白葡萄酒，鹽、白胡椒醃製，再沾上麵粉→蛋液→麵包粉炸成金黃色。

② 將炸好的蚵仔，放入裝飾材料，拌勻，並放上法式麵包上即可。

Methods：

1. Marinade oyster with white wine、salt and pepper. Breading with flour、egg and breadcrumb and deep fry until golden brown.

2. Combine cooked oyster and garnish items. Mix well. And put all of the ingredients on French bread. Serve as need.

38

酥炸雞腿三明治
Fried Chicken leg Sandwich

主要材料：

去骨雞腿肉........1顆
法式芥茉醬..10毫升
鹽、白胡椒......適量
麵粉............50公克

麵包：

漢堡包..............1顆

抹醬：

軟化奶油......15公克

裝飾材料：

美生菜..............1片
蜜洋蔥........20公克

做法：

① 將去骨雞腿放入法式芥茉醬及鹽、白胡椒醃製約半小時，再沾上麵包炸成金黃色即可。

② 將漢堡包切開抹上奶油，並烤上色，放上裝飾材料並放上炸好的雞腿即可。

Ingredients
Body：
Chicken leg
Boneless 1EA
French mustard
10ml
S&P TO TASTE
Flour 50gr

Garnish：
Green lettuce 1p.c
Onion comfit 20gr

Bread：
Burger Bun 1EA

Spread：
Melted Butter 15gr

Methods：

1. Marinade boneless chicken leg with French mustard、salt and pepper for 30 minutes. Breading and deep fry until the color go to golden brown.

2. Cut open burger bun and spread butter on cutting surface, toast until colored. Put on garnish items and cooked chicken legs. Serve as need.

↓廣東A菜(綠萵苣) Cantonese
又稱劍葉萵苣，為台灣本土很常見的蔬菜，富含很多營養素，可清炒、煮湯。因植株本身帶有一種特殊的風味，較不易被蟲類啃食，相對的農藥的用量也很少，可放心食用。

Open Face Sandwich

開面三明治

燻鮭魚開面三明治 39

Smoked Salmon and Onion Ring on French bread

主要材料：
燻鮭魚 8片

裝飾材料：
生菜葉 8片
洋蔥圈 1顆

麵包：
法國麵包.......................1顆

抹醬：
美奶滋.....................20毫升
狄蓉芥茉醬..............10毫升

做法：
將法國麵包切開抹上芥茉美
奶滋，放上主要及裝飾材料
即可。

Ingredients
Body：
Smoked Salmon Sliced 8p.c

Garnish：
Green lettuce 8p.c
Onion Ring 1EA

Bread：
French bread 1EA

Spread：
Mayonnaise 20ml
Dijon mustard 10ml

Methods：
Slice French bread. Spread
mayonnaise and mustard
on bread. Put body and
garnish items on top. Serve
as need.

鮪魚開面三明治
Tuna Fish Salad on Brown Toast

主要材料：
鮪魚罐 1罐

裝飾材料：
A ⌈ 洋蔥碎 50公克
  │ 西芹碎 30公克
  │ 酸黃瓜碎........30公克
  │ 巴西里碎.........5公克
  └ 鹽、白胡椒 ....適量
蕃茄片 4片
水煮蛋片 1顆

麵包：
全麥吐司.................4片

抹醬：
美奶滋................80毫升

做法：
① 將主要材料和裝飾材料A拌勻成鮪魚沙拉。
② 蕃茄片、水煮蛋片及揉球狀的鮪魚沙拉放於麵包上即可。

白吐司/全麥吐司/厚片吐司
White Toast / Brown Toast
在世界各地相當普遍的麵包種類，食用方式相當多，可烤、炸、煎或是直接食用，白吐司主要是由普通麵粉製成；全麥吐司是由全麥麵粉(Whole wheat flour)製成。

Ingredients
Body：
Tuna fish canned 1Tin

Garnish：
A ⌈ Onion chopped 50gr
  │ Celery chopped 30gr
  │ Pickle chopped 30gr
  │ Parsley chopped 5gr
  └ S&P TO TASTE
Tomato Slice 4p.c
Hard Boiled Egg Slice 1EA

Bread：
Brown Toast 4p.c

Spread：
Mayonnaise 80ml

Methods：
1. Combine body and garnish A items and mix well.
2. Put tomato slice,hard boiled egg slice and shape the tuna salad into ball shape and place on brown toast. Serve as need.

烏魚子開面三明治
Mullet Fish Roe with Leek on White Toast

主要材料：
烏魚子 30公克
白蘭地 30毫升

裝飾材料：
白蘿蔔片 10片
青蒜苗 15公克

麵包：
白吐司......................2片

抹醬：
美奶滋 30毫升

做法：
① 將烏魚子泡白蘭地再烤約3分鐘，切片備用。
② 將白吐司抹上美奶滋切開備用。
③ 將白蘿蔔切片，青蒜苗切片，擺上白吐司上即可。

Ingredients
Body：
Mullet Fish Roe 30gr
Brandy 30ml

Garnish：
Turnips sliced 10p.c
Leek 10gr

Bread：
White Toast 2p.c

Spread：
Mayonnaise 30ml

Methods：
1. Soak fish roe in brandy for a while, then roast in oven for 3 minutes, slice for use.
2. Spread mayonnaise on white toast, cut for use.
3. Slice turnips and leek, place roe、turnip and leek slices on toast. Serve as need.

鳳梨火腿開面三明治
Ham and Pineapple Sliced on White Toast

主要材料：
火腿 4片

裝飾材料：
鳳梨片 1/4顆
藍莓 8顆

麵包：
白吐司......................4片

抹醬：
美奶滋 20毫升

做法：
將白吐司切成小長方，抹上美奶滋，並放上主要及裝飾材料即可。

Ingredients
Body：
Ham 4p.c

Garnish：
Pineapple sliced 1/4EA
Blueberry 8EA

Bread：
White Toast 4p.c

Spread：
Mayonnaise 20ml

Methods：
Cut white toast into small square, spread mayonnaise on top. Put on body and garnish items. Serve as need

43 火腿蛋開面三明治

Hard Boiled Egg with Ham on white Toast sliced

主要材料：
蛋片.................2顆
火腿片.............2片

裝飾材料：
小黃瓜切絲......適量
覆盆子.............8顆

麵包：
白吐司.............2片

抹醬：
美奶滋.........30毫升

做法：
① 白吐司抹上美奶滋切塊備用。
② 蛋煮熟切片，火腿切片，小黃瓜切絲，擺
　上切好的白吐司即可。

Ingredients
Body：
Sliced Hard Boiled Egg
2EA
Ham sliced 2p.c

Garnish：
Baby Cucumber Julienne
as need
Raspberry 8EA

Bread：
White Toast 2p.c

Spread：
Mayonnaise 30ml

Methods：
1. Spread mayonnaise on white toast and cut for use.
2. Slice boiled egg,ham,raspberry and baby cucumber,
   placesliced ingredients on top of white toast.
   Serve asneed.

花枝圈開面三明治 **44**

Squid Ring and Onion Comfit on French bread

主要材料：
花枝圈 1個
蜜洋蔥.....................30公克

裝飾材料：
義式甜醋汁..............20毫升
洋蔥絲.....................10公克

麵包：
法國麵包片 6片

抹醬：
軟化奶油 30毫升

做法：
將麵包斜切抹上軟化奶油放
上主要及裝飾材料即可。

Ingredients
Body：
Squid Ring 1EA
Onion comfit 30gr

Garnish：
Balsamic Dressing
20ml
Onion Julienne 10gr

Bread：
French bread Sliced
6EA

Spread：
Melted Butter 30ml

Methods：
Slice French bread and
spread melted butter. Put
body and garnish ingredi-
ents on top. Serve as need.

45

鮮蝦蕃茄開面三明治
Shrimp and Tomato Sliced on French bread

主要材料：
草蝦.................12尾

麵包：
法國麵包...........1顆

裝飾材料：
生菜葉...............3片
小黃瓜切片........1條

抹醬：
美奶滋.........30毫升
日式芥茉醬..10公克

做法：
將法國麵包切片抹上芥茉美奶滋，再放上主要
及裝飾材料即可。

Ingredients
Body：
Shrimp 12EA

Garnish：
Green Lettuce 3p.c
Baby Cucumber sliced
1EA

Bread：
French bread 1EA

Spread：
Mayonnaise 30ml
Wasabi 10gr

Methods：
Slice French bread. Spread mayonnaise and Wasabi
on bread. Put body and garnish items on top. Serve
as need.

黃瓜沙拉開面三明治

Dill Cucumber Salad on French bread

主要材料：
大黃瓜切片..........1顆
鹽 5公克
煙燻火雞肉片 50公克

裝飾材料：
優格............250毫升
新鮮蒔蘿.........5公克
生菜葉...............5片

麵包：
法國麵包.............1顆

抹醬：
美奶滋..........20毫升

做法：

① 將大黃瓜切片，放入鹽軟化待出水擠乾水份
　 加入裝飾材料拌勻。

② 將拌勻的黃瓜沙拉及煙燻火雞肉片放於法國
　 麵包上即可。

Ingredients
Body：
Cucumber sliced 1EA
Salt 5gr
Smoked Turkey Ham 50gr

Garnish：
Yoghurt 250ml
Dill Fresh 5gr
Green lettuce 5p.c

Bread：
French bread 1EA

Spread：
Mayonnaise 20ml

Methods：

1. Slice the cucumber. Add salt to soft the cucumber. Squeeze out of the water of cucumber. Mix with garnish items.

2. Place cucumber salad and smoked turkey ham on French bread. Serve as need.

華爾道夫沙拉
開面三明治

Waldorf salad on White Toast

主要材料：
蘋果丁.................................1顆
西芹丁.........................60公克
葡萄乾.........................20公克

裝飾材料：
核桃.............................20公克
西芹葉.............................5公克
鹽、白胡椒.......................適量

麵包：
漢堡包.............................4個

抹醬：
美奶滋.........................80毫升

做法：
① 將主要材料與裝飾材料及
　 美奶滋拌勻。
② 放於漢堡包內即可。

Ingredients
Body：
Apple diced 1EA
Celery diced 60gr
Raisins 20gr

Garnish：
Walnut 20gr
Celery leaves 5gr
S&P TO TASTE

Bread：
Hamburger Bun 4EA

Spread：
Mayonnaise 80ml

Methods：
1. Combine body and garnish items. Add mayonnaise and mix well.
2. Place Waldorf salad in hamburger bun. Serve as need.

水蜜桃起士開面三明治
Cream Cheese and Peach sliced on Walnut Bread

主要材料：
奶油起士....120公克
蜂蜜.............10毫升

裝飾材料：
水蜜桃片 1顆
核桃碎 ....適量

麵包：
核桃麵包 1顆

抹醬：
美奶滋 50毫升

做法：
將奶油起士加入蜂蜜調勻，放上水蜜桃片再放於核桃麵包上即可。

Ingredients
Body：
Cream cheese 120gr
Honey 10ml

Garnish：
Peach sliced 1EA
Walnut Chopped as need

Bread：
Walnut Bread 1EA

Spread：
Mayonnaise 50ml

Methods：
Mix cream cheese and honey. Put peach slices on walnut bread. And put the mixture of body items on top. Serve as need.

核桃麵包Walnut bread
由糖、油脂、蛋、胡桃和全麥麵粉所製成的麵包，具有十足的堅果香味。

49

火腿起士開面三明治
Ham and Cheese Sandwich

主要材料：
火腿.................4片
葛利亞起士........4片
無花果..............2顆

麵包：
白吐司...............4片

抹醬：
美奶滋.........50毫升

做法：
將白吐司切好抹上美奶滋，再擺上火腿、起士及無花果即可。

Ingredients
Body：
Ham 4p.c
Gruyere cheese 4p.c
Fig Slice 2EA

Bread：
White Toast 4p.c

Spread：
Mayonnaise 50ml

Methods：
Spread mayonnaise on white toast. Then put ham , gruyere cheese and fig slice on top. Serve as need.

無花果 Fig
主要出產自溫帶氣候區，形狀多為橢圓狀或梨形，具有厚實的果皮，果肉為褐紫色且分佈許多細小的籽，市面上可買到新鮮產品，也買得到可長時間保存的乾燥果實。

焗蔬菜丁起士開面三明治

Gratinated Vegetable diced and Mozzarella cheese Sandwich

主要材料：

青甜椒丁 30公克
紅甜椒丁 30公克
黃甜椒丁 30公克
洋蔥丁 30公克
小黃瓜丁 30公克
茄子丁 30公克

裝飾材料：

莫札瑞拉起士50公克
披薩草............3公克
橄欖油..........15毫升

麵包：

法國麵包 1個

抹醬：

軟化奶油......20毫升

做法：

① 將法國麵包斜切片，抹上奶油烤上色備用。

② 將主要材料用橄欖油炒熟，擺放在烤好的麵
包上，再放上莫札瑞拉起士焗上色即可。

Ingredients
Body：
Green pepper diced 30gr
Red pepper diced 30gr
Yellow pepper diced 30gr
Onion diced 30gr
Baby cucumber diced 30gr
Egg plant diced 30gr

Garnish：
Mozzarella cheese 50gr
Oregano 3gr
Olive oil 15ml

Bread：
French bread 1EA

Spread：
Melted Butter 20ml

Methods：
1. Slice French bread, spread butter and toast until golden brown.
2. Sauté body items with olive oil. Make sure that all the ingredients are cooked. Then put cooked vegetables on French bread.
3. Put mozzarella cheese on top of vegetables. And gratin until colored. Serve as need.

51

鮮菇牛肉開面三明治
Sautéed Beef Sliced and Mushroom Sandwich

主要材料：
腓力牛肉片 150公克

麵包：
粿麥麵包 1顆

裝飾材料：
洋菇 30公克
香菇 30公克
鮑魚菇 30公克
洋蔥 30公克

抹醬：
軟化奶油 20毫升

做法：
① 粿麥麵包抹上軟化奶油。
② 將腓力切片，與裝飾材料炒熟。
③ 放上粿麥麵包即可上桌。

Ingredients
Body：
Beef Tenderloin Sliced
150gr

Garnish：
Mushroom button 30gr
Mushroom shitake 30gr
Mushroom abalone
30gr
Onion Sliced 30gr

Bread：
Rye Bread 1EA

Spread：
Melted Butter 20ml

Methods：
1. Spread butter on rye bread.
2. Slice beef fillet and sauté with garnish items.
3. Place cooked ingredients on top of bread. Serve as need.

烤杏鮑菇開面三明治

Grilled Scallops Mushroom Sandwich

主要材料：
杏鮑菇 120公克
蘆筍尖 50公克

裝飾材料：
大蒜碎 5公克
橄欖油 15毫升
鹽、白胡椒

麵包：
全麥麵包 4片

抹醬：
美奶滋 30毫升
狄蓉芥茉醬 ..30毫升

做法：
① 將全麥麵包切好抹上美奶滋。
② 將杏鮑菇切片，抹上大蒜碎、橄欖油烤熟，蘆筍尖煮熟放上麵包，擠上狄蓉芥茉醬即可。

Ingredients
Body：
Scallops mushroom 120gr
Asparagus Tip 50gr

Garnish：
Garlic chopped 5gr
Olive oil 15ml
S&P TO TASTE

Bread：
Brown Toast 4p.c

Spread：
Mayonnaise 30ml
Dijon Mustard 30ml

Methods：
1. Slice brown toast and spread mayonnaise on bread.
2. Slice scallop mushrooms, sprinkle garlic chopped and olive oil on top. Roast in oven. Cooked the asparagus with hot water.
3. Put mushrooms and asparagus on bread add dijon mustard, serve as need.

芥茉粉 Mustard Powder
質地細緻，為銘黃色的粉末，和其他同類調味料相較它具有較柔和的辛辣味，可在市面上買到罐裝的成品。

53

義式香腸法士達

Focaccia Topping with Salami and mozzarella Cheese

主要材料：
義式香腸 50公克
火腿 20公克

裝飾材料：
洋蔥絲 30公克
紅甜椒 30公克
黃甜椒 30公克
莫札瑞拉起士 50公克

麵包：
法士達麵包 1顆

抹醬：
橄欖油 5毫升

做法：
將法士達麵包淋抹上橄欖油放上主要與裝飾材料，烤上色即可。

Ingredients
Body：
Salami 50gr
Ham 20gr

Garnish：
Onion Sliced 30gr
Red pepper 30gr
Yellow pepper 30gr
Mozzarella cheese 50gr

Bread：
Focaccia 1EA

Spread：
Olive oil 5ml

Methods：
Sprinkle some olive oil Focaccia, put on body and garnish items on top. Roast in oven until colored. Serve as need.

莫札瑞拉起士
Mozzarella Cheese
義大利南方的一種乳酪，原來是使用水牛奶來釀造，但現今大都以牛奶來製作，質地白且味道溫和。

奶油起士 Cream Cheese
以牛奶鮮奶油和牛奶混合物製成的一種新鮮、柔軟、溫和、白色的乳酪，適合用在烘焙、麵包塗抹、糖果等用途。通常以塊狀販售。

鴨肉吐司披薩 54

White Toast Topping with Roasted Duck and Mozzarella Cheese

主要材料：
烤鴨肉.........................120公克

裝飾材料：
洋蔥絲.........................50公克
莫札瑞拉起士...............50公克
披薩草............................1公克

麵包：
白吐司..............................1顆

抹醬：
軟化奶油......................30公克

做法：
將白吐司抹上奶油，放上主要
及裝飾材料烤上色即可。

Ingredients
Body：
Roasted Duck Sliced
120gr

Garnish：
Onion Sliced 50gr
Mozzarella cheese 50gr
Oregano 1gr

Bread：
White Toast 1EA

Spread：
Melted Butter 30gr

Methods：
Spread butter on white toast. Put body and garnish
items on top. Roast in oven until colored.

55

水果吐司披薩

White Toast Topping with Fresh Fruit and Mozzarella cheese

主要材料：

草莓 5顆
奇異果 1顆
哈蜜瓜 120公克
鳳梨 120公克

裝飾材料：

萊姆酒 10毫升
糖 50公克
莫札瑞拉起士 50公克

麵包：

白吐司................................2片

抹醬：

花生醬 20公克

做法：

① 將白吐司抹上花生醬備用。

② 將主要材料切塊拌上萊姆酒
及糖，再放上麵包上再放上
莫札瑞拉起士烤上色即可

Ingredients
Body：
Strawberry 5EA
Kiwi 1EA
Honey dew melon 120gr
Pineapple sliced 120gr

Garnish：
Rum 10ml
Sugar 50gr
Mozzarella cheese 50gr

Bread：
White Toast 2EA

Spread：
Peanut Butter 20gr

Methods：
1. Spread peanut butter on white toast. Cut body items into dices. And mix with rum and sugar.
2. Put fruits dices on toast, and put on mozzarella cheese, then finish in oven until colored.

Staffed and Rolled

Sandwich

填入 ＆ 包捲三明治

56 蛋沙拉填入三明治

Hard Boiled Egg Salad in Baguette

主要材料：
水煮蛋 5顆
芥茉粉........................5公克
青蔥 15公克
美奶滋....................80毫升

裝飾材料：
黑橄欖 5顆
酸黃瓜 15公克
法國麵包....................1/2顆

做法：
① 將法國麵包挖空備用。
② 將主要材料拌勻和裝飾材料一起填入挖空的法國麵包內塞緊後，再切片即可。

Ingredients
Body：
Hard Boiled Egg
chopped 5EA
Mustard powder 5gr
Spring onion chopped
15gr
Mayonnaise 80ml

Garnish：
Black olive 5EA
Pickle 15gr

Bread：
Baguette 1/2EA

Methods：
1. Hollow out a baguette for use.
2. Combine all the ingredients of body items and mix well. Stuff the mixture and garnish items into the hollow bread. Make sure the structure is firm.
3. Slice and serve.

鮪魚填入三明治

Tuna fish salad in Baguettes

主要材料：

鮪魚罐..............1罐
洋蔥碎........50公克
西芹碎........30公克
酸黃瓜碎......30公克
巴西里碎........5公克
鹽、白胡椒......適量
美奶滋.......120毫升

裝飾材料：

蕃茄...............1/2顆
橄欖油..............5顆
水煮蛋..............1顆

麵包：

法國麵包........1/2顆

做法：

① 將法國麵包挖空備用。

② 將主要材料拌勻和裝飾材料一起填入挖空的法國麵包內塞緊。

③ 再用紙包好切成兩半即可。

Ingredients

Body：

Tuna fish canned 1Tin
Onion chopped 50gr
Celery chopped 30gr
Pickle chopped 30gr
Parsley chopped 5gr
S&P TO TASTE
Mayonnaise 120ml

Garnish：

Tomato diced 1/2EA
Black Olive 5EA
Hard Boiled Egg 1EA

Bread：

Baguettes 1EA

Methods：

1. Hollow out a baguette for use.
2. Combine all the ingredients of body items and mix well. Stuff the mixture and garnish items into the hollow bread. Make sure the structure is firm.
3. Wrap with a piece of paper, cut into half. Serve as need.

酥炸吐司杯

Fried Toast Cup Stuffed with Tuna fish Salad and Hard Boiled Salad

主要材料：
鮪魚沙拉 ...120公克(→p.10)
蛋沙拉 ......120公克(→p.10)

裝飾材料：
生菜葉 6片
火腿片.........................6片

麵包：
未切片白吐司.............1/2條

做法：
① 將白吐司切成正方體挖空成杯狀，並炸成金黃色。
② 將主要材料及裝飾材料填入裝飾即可。

Ingredients
Body：
Tuna fish salad 120gr
(→p.10)
Hard Boiled Egg salad
120gr(→p.10)

Garnish：
Lettuce leaves 6p.c
Ham sliced 6p.c

Bread：
White Toast Half loaf

Methods：
1. Cut white toast into Small Square. Hollow out the white toast and make it into a cup shape. Deep fried until golden brown.
2. Put body and garnish items into the cup. Serve as need.

牛肉火腿片 Beef Ham
將牛肉醃漬(有些產品會經過醃燻的過程)後，真空保裝出售。有時可買到切片的包裝。

59

希臘式沙拉口袋餅 Greek Salad in Pita

主要材料：
羊乳酪...........................80公克
羅蔓生菜.......................50公克
比利時生菜...................30公克
紫包心菜.......................30公克
捲鬚生菜.......................10公克
洋蔥絲...........................20公克

裝飾材料：
黑橄欖............................5顆
義式甜醋汁...................20毫升

麵包：
口袋餅..............................1顆

做法：
① 將口袋餅切成兩半備用。
② 將主要材料填入口袋餅再淋
　 上裝飾材料即可。

Ingredients
Body：
Feta cheese 80gr
Romaine lettuce 50gr
Endive lettuce 30gr
Reddish lettuce 30gr
Fresee lettuce 10gr
Onion Sliced 20gr

Garnish：
Black olive 5EA
Balsamic Dressing 20ml

Bread：
Pita 1EA

Methods：
1. Cut pita into a half.
2. Put body items into pita
   bag, sprinkle with garnish
   dressing. Serve as need.

羊起士 Feta Cheese
是以羊乳(也有使用山羊乳)製成
的，白色且質地易碎，酸且鹹的風
味相當強烈，市面上可購買到泡在
鹽水的罐裝產品。

炒豬肉口袋餅 60

Sautéed Pork Sliced With Mushroom in Pita

主要材料：
豬肉片 120公克

裝飾材料：
洋蔥絲 50公克
洋菇片 80公克
百里香 1公克
玉桂葉 1片
白葡萄酒 30毫升
鮮奶油 80毫升
鹽、胡椒 ....適量
生菜葉 2片

麵包：
口袋餅 1個

做法：
① 將口袋餅切開成兩半備用。
② 將主要材料及裝飾材料炒熟再放入口袋餅即可。

口袋餅 Pita
屬於中東地區的麵包，以酵母發酵膨脹來撐起開口，通常可切開呈袋狀後放入餡料享用。

Ingredients
Body：
Pork meat sliced 120gr

Garnish：
Onion Sliced 50gr
Mushroom Button sliced 80gr
Thyme 1gr

Bay leaf 1p.c
White wine 30ml
Cream 80ml
S&P TO TASTE
Green lettuce 2p.c

Bread：
Pita 1EA

Methods：
1. Cut pita into a half.
2. Sauté body and garnish items until cooked. Put cooked ingredients into pita bag. Serve as need.

61 炸花枝圈三明治 Submarine

主要材料：
花枝圈.........3~4片
蛋 1顆
麵粉...............少許
麵包粉............少許
鹽、胡椒粉.....適量

裝飾材料：
蘿蔓生菜心........1個

做法：

① 以鹽和胡椒調味花枝圈,依序沾裹麵粉、蛋液和麵包粉後,炸至金黃色後備用。

② 將炸好的花枝圈層層疊起,中間放入蘿蔓生菜心裝飾即可。

Ingredients
Body：
Cuttlefish rings 3~4p.c
Egg 1EA
Flour Few
Breadcrumb Few
S&P To taste

Garnish：
Romaine lettuce heart
1EA

Methods：

1. Marinade cuttlefish rings with S&P. Breading the cuttlefish rings with flour、egg and breadcrumb. Deep fry until golden brown.

2. Layer the cuttlefish rings and put romaine heart in the middle of cuttlefish rings for garnish. Serve as need.

煙燻鮭魚捲 Smoked Salmon and Cream Cheese Roll

主要材料：
燻鮭魚..............6片

麵包：
墨西哥玉米餅....1片

裝飾材料：
美生菜.........50公克
洋蔥絲.........50公克

抹醬：
奶油起士....120公克

Ingredients
Body：
Smoked Salmon 6p.c

Garnish：
Iceberg lettuce Sliced 50gr
Onion Sliced 50gr

Bread：
Tortilla 1EA

Spread：
Cream cheese 120gr

做法：
將墨西哥玉米餅先抹上奶油起士，再撒上美生菜及洋蔥絲，再擺上燻鮭魚捲起來，再切片即可。

Methods：
Spread cream cheese on tortilla, sprinkle with lettuce and onion slices. Place smoked salmon on top and roll up. Slice and serve.

墨西哥鮭魚捲
Smoked Salmon in Tortilla

主要材料：　　　　麵包：
燻鮭魚..............5片　墨西哥玉米餅....1片
洋蔥絲.........35公克
蕃茄沙沙醬..50公克　抹醬：
美生菜.........50公克　酪梨醬.........50公克

做法：
將墨西哥玉米餅抹上酪梨醬再放上主要及裝飾
材料捲起，再切即可。

Ingredients
Body：
Smoked salmon
5p.c

Bread：
Tortilla 1EA

Spread：
Guacamole 50gr

Garnish：
Onion Sliced 35gr
Tomato salsa 50gr
Iceberg lettuce
50gr

Methods：
Spread guacamole on tortilla, Put body and
garnish items on tortilla, roll up. Slice and
serve.

墨西哥玉米餅 Tortilla
以玉米麵團或小麥粉及油
脂拌勻煎烤而成。形狀圓
薄且無膨脹。適合搭配餡
料一起享用。

墨西哥雞肉捲 64

Grilled Chicken Breast and Tomato Salsa in Tortilla

主要材料：

雞胸肉.........................120公克

裝飾材料：

洋蔥絲.........................30公克
美生菜.........................50公克
哈蜜瓜.........................50公克
蕃茄沙沙醬...................50公克

麵包：

墨西哥玉米餅.....................1片

抹醬：

酪梨醬.........................50公克

做法：

① 將雞胸肉烤熟切條備用。

② 將墨西哥玉米餅抹上酪梨醬再放上主要及裝飾材料捲再切開即可。

Ingredients
Body：
Chicken Breast 120gr

Garnish：
Onion Sliced 30gr
Iceberg lettuce 50gr
Honey dew melon 50gr
Tomato salsa 50gr

Bread：
Tortilla 1EA

Spread：
Guacamole 50gr

Methods：

1. Grill chicken breast and cut into strips for use.
2. Spread guacamole on tortilla, put all of the body and garnish ingredients on tortilla. Roll up. Slice and serve.

65 炒牛蔬菜口袋餅

Sauté Beef Sliced and Bell Peppers in Pita

主要材料：
牛腓力切片 120公克

裝飾材料：
青甜椒絲......50公克
黃甜椒絲......50公克
紅甜椒絲......50公克
洋蔥絲 50公克
橄欖油 15毫升
鹽、黑胡椒 ......適量

麵包：
口袋餅 1個

Ingredients
Body：
Beef Tenderloin sliced
120gr

Garnish：
Green pepper sliced
50gr
Yellow pepper sliced
50gr
Red pepper sliced 50gr
Onion sliced 50gr
Olive oil 15ml
S&P TO TASTE

Bread：
Pita 1EA

做法：
① 將口袋餅切成兩半
   備用。
② 將主要材料及裝飾
   材料炒熟，放入口
   袋餅即可。

Methods：
1. Cut pita into a half.
2. Sauté body and garnish items until cooked. Put cooked ingredients into pita bag. Serve as need.

66

鮪魚脆餅桶
Stuffed Tuna fish in Crispy Cracker

主要材料：
鮪魚罐......................1罐

裝飾材料：
洋蔥碎 50公克
西芹碎 30公克
酸黃瓜碎...........30公克
巴西里碎...............5公克
鹽、白胡椒............適量
美奶滋 80毫升

麵包：
脆餅桶....................5個

做法：
將主要材料及裝飾材料拌勻，用冰淇淋杓挖球放入脆餅桶即可。

Ingredients
Body：
Tuna fish canned 1Tin

Garnish：
Onion chopped 50gr
Celery chopped 30gr
Pickle chopped 30gr
Parsley chopped 5gr
S&P TO TASTE
Mayonnaise 80ml

Bread：
Crispy Cracker 5EA

Methods：
Combine all of the body and garnish items. Mix well. Use ice cream scoop to make salad into ball shape. Then place in crispy cracker. Serve as need.

67

蛋沙拉杯　Hard Boiled Egg Salad in Crispy Cracker Bowl

主要材料：
水煮蛋沙拉 120公克
（→p.18）

裝飾材料：
小蕃茄..............5顆
洋芋球..............5顆
哈蜜瓜球.........5顆
黑橄欖..............5顆

麵包：
脆餅杯..............1個

做法：
將水煮蛋沙拉用冰淇淋挖
成小球與裝飾材料放入脆
餅杯即可。

Ingredients
Body：
Hard Boiled Egg Salad
120gr（→p.18）

Garnish：
Cherry Tomato 5EA
Boiled potato Ball 5EA
Honey dew melon Ball
5EA
Black Olive 5EA

Bread：
Crispy Cracker Bowl
1EA

Methods：
Use ice cream scoop to
shape egg salad into ball
shape. Put egg salad
and garnish items in to
crispy cracker bowl.
Serve as need.

南美醋栗Cape gooseberry
是一種櫻桃般大小的黃橙色
水果，原產於南美洲。其外
殼薄如紙張，如燈籠般包覆
整顆漿果，果肉呈現乳白色
且多細籽，風味特殊，因其
外殼特殊的造型很適合當菜
餚的裝飾，亦可做烘焙食材
或是果醬。

Pancake

煎餅 · 鬆餅

Waffle

煎餅・鬆餅 基本做法

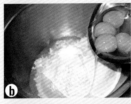

共同做法：

a 麵粉過篩和糖粉一起放入鋼盆。

b 將蛋加入拌勻。

c 倒入牛奶混合。

d 再加入蜂蜜、水及香草精。

e 將材料混合均勻成麵糊。

材料：

低筋麵粉 200公克
糖粉......................90公克
蛋..........................4顆
香草精..................5毫升
牛奶....................120毫升
沙拉油..................40毫升
蜂蜜.....................35毫升
泡打粉....................10克

Ingredients

Cake Flour 200g
Icing sugar 100g
Egg 200ml   Salad oil 40ml
Vanilla Extract 5ml Honey 35ml
Milk 120ml   B.P. 10g

Methods :

1. Sieve flour and icing sugar, combine with eggs.
2. Add milk and mix well.
3. Combine with honey、 water and vanilla extract, mix well.

做法：

煎餅

① 取一勺麵糊倒入鋪成薄圓片狀。

Pour one ladle of batter and pour on top of pan.

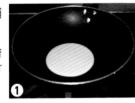

② 待麵糊加熱至體積澎脹起泡泡。

Cook until batter puff up and bubble shown on surface.

③ 即翻面，兩面煎至金黃色即可取出。

Turn it over. Cook until golden brown. Serve as need.

鬆餅

做法：

① 取適量麵糊放入鬆餅機中（鬆餅機應完成加熱步驟）。

Put good quantity of batter in waffle iron.

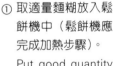

② 蓋上鬆餅機。

Cover the machine.

③ 待機器亮燈，鬆餅呈金黃色即可。

Wait for a few minutes until machine is done. Make sure waffle is golden brown.

烤鴨煎餅

Roasted Duck and onion comfit on Pancake

主要材料：
烤鴨..........120公克

麵包：
煎餅.................2片

裝飾材料：
蜜洋蔥.........50公克
生菜葉..............2片
香菜..................1支

抹醬：
梅醬.............20毫升

做法：
① 先煎好煎餅。
② 再抹上梅醬，再擺上主要材料及裝飾材料即可。

Ingredients
Body：
Roasted Duck Sliced
120gr

Garnish：
Onion comfit 50gr
Green lettuce 2p.c
Coriander 1EA

Bread：
Pancake 2p.c

Spread：
Plum Sauce 20ml

Methods：
1. Cook pancake in pan until golden brown.
2. Spread plum sauce on pancake. And put body and garnish items on top. Serve as need.

燻鮭魚煎餅

Smoked Salmon and Sour Cream Pancake

主要材料：
燻鮭魚..........................5片

裝飾材料：
洋蔥絲......................30公克
生菜絲......................60公克

麵包：
煎餅................................2片

抹醬：
酸奶油.....................50毫升

做法：
① 先將煎餅煎好。
② 抹上酸奶油再放上主要及
　 裝飾材料即可。

Ingredients
Body：
Smoked Salmon 5p.c

Garnish：
Onion Sliced 30gr
Green lettuce Julienne
60gr

Bread：
Pancake 2p.c

Spread：
Sour Cream 50ml

Methods：
1. Cook pancake in pan until golden brown.
2. Spread sour cream on pancake. And put body and garnish items on top. Serve as need.

煙燻鮭魚 Smoked Salmon
以熱燻或是冷燻方式所做出鮭魚，
通常以挪威或是北歐國家產的鮭魚
為原料，包裝上通常會寫上出產國
和添加的原料。口感帶有鹹味，可
搭配酸豆、洋蔥等來直接食用，亦
可加熱烹調後食用。

70

炒鮮菇煎餅
Sautéed Mixed Mushroom on Pancake

主要材料：
洋菇............30公克
香菇............30公克
杏鮑菇 30公克
鮑魚菇 30公克

裝飾材料：
橄欖油 15毫升
洋蔥塊 50公克
大蒜 5公克
羅勒 5公克

麵包：
煎餅 2片

抹醬：
軟化奶油......20毫升

做法：
先將煎餅煎好，抹上奶油，再將主要及裝飾材料炒好放上即可。

Ingredients
Body：
Mushroom Button 30gr
Mushroom Shitake 30gr
Mushroom Scallop 30gr
Mushroom Abalone 30gr

Garnish：
Olive oil 15ml
Onion diced 50gr
Garlic diced 5gr
Basil 5gr

Bread：
Pancake 2p.c

Spread：
Melted Butter 20gr

Methods：
1. Cook pancake in pan until golden brown.
2. Spread butter on pancake. Sauté body and garnish items until cooked.
3. Put cooked ingredients on top. Serve as need.

葛利亞起士片 Gruyere Cheese
為瑞士產的一種乳酪，以牛奶釀製，乳酪表面為褐金黃色，內蕊部分為淡黃色，乳酪內部有大孔洞，味道濃郁、甘甜且帶有些許堅果芳香味。

藍乳酪 Blue Cheese
具有藍綠色黴菌的乳酪，具有極強烈的酸和辛辣味，還具有一股特殊的香味。美、加、法等各國都有生產類似的產品，其中以法國產的Roquefort藍紋乳酪最具代表性。

水果冰淇淋煎餅 71

Fresh Fruit and Ice Cream on Pancake

主要材料：
冰淇淋 1球

裝飾材料：
草莓 3顆
楊桃 1顆
哈蜜瓜 80公克
香蕉 80公克
奇異果 1顆
萊姆酒 20毫升
糖 40公克
檸檬皮碎 1顆

麵包：
煎餅 1片

抹醬：
打發鮮奶油80毫升

做法：
① 先將煎餅煎好。
② 將裝飾材料拌勻後放在煎餅上，再放上喜愛的冰淇淋即可。

Ingredients
Body：
Ice Cream (your choice)
1Scoop

Garnish：
Strawberry 3EA
Star Fruit 1EA
Honey dew melon 80gr
Banana 80gr
Kiwi 1EA

Rum 20ml
Sugar 40gr
Lemon zest chopped
1EA

Bread：
Pancake 1p.c

Spread：
Whipped Cream 80ml

Methods：
1. Cook pancake in pan until golden brown.
2. Mix all of the garnish items. Put garnish items on pancake and serve with your favorite ice cream.

蟹肉沙拉煎餅

Crab Meat Salad on Pancake

主要材料：
蟹肉..........120公克

麵包：
煎餅..................2片

裝飾材料：
洋蔥絲 30公克
綜合生菜葉 120公克
義式甜醋汁 ..30毫升

抹醬：
軟化奶油......20毫升

做法：
先將煎餅煎好抹上奶油，再將煮熟的蟹肉與
裝飾材料拌勻，再放上煎餅即可。

Ingredients
Body：
Crab Meat 120gr

Garnish：
Onion Sliced 30gr
Mixed Green 120gr
Balsamic Dressing
30ml

Bread：
Pancake 2p.c

Spread：
Melted Butter 20gr

Methods：
1. Cook pancake in pan until golden brown.
2. Mix cooked crab meat with garnish items.
3. Place crab meat salad on top of pancake. Serve as need.

千層蕃茄蛋煎餅

Egg Salad and Tomato Tower with Pancake

主要材料：
水煮蛋沙拉 150公克

裝飾材料：
蕃茄..................1顆
生菜葉..............2片

做法：
先煎好煎餅，再將水
煮蛋沙拉與裝飾材料
及煎餅一層一層排好
即可。

Ingredients
Body：
Egg Salad 150gr

Garnish：
Tomato sliced 1EA
Green lettuce 2p.c

Bread：
Pancake 2p.c

Methods：
1. Cook pancake in pan until golden brown.
2. Layer pancake、body and garnish items, and shape all ingredients into a tower. Serve as need.

苦菊苣 Belgian Endive
又名比利時菊苣，體型長類似小
型包心菜，葉尖端呈黃色,葉片呈
乳色，味道微苦。通常使用在沙
拉中。

水果沙拉煎餅捲
Fresh Fruit Salad in Pancake Roll

主要材料：

草莓 3顆
楊桃 1顆
哈蜜瓜 80公克
香蕉.........................80公克
奇異果.........................1顆
萊姆酒 20毫升
檸檬皮碎 1顆
糖 40公克
藍莓 8顆

裝飾材料：

糖粉.........................30公克

麵包：

煎餅 1片

做法：

① 先將煎餅煎好。
② 將主要材料拌勻，放入煎
餅捲內再灑上糖粉即可。

Ingredients
Body：
Strawberry 3EA
Star Fruit 1EA
Honey dew melon 80gr
Banana 80gr
Kiwi 1EA
Rum 20ml
Lemon zest chopped 1EA
Sugar 40gr
Blueberry 8EA

Garnish：
Icing Sugar 30gr

Bread：
Pancake 1p.c

Methods：

1. Cook pancake in pan until golden brown.
2. Mix all of the body ingredients, and put into the pancake roll. Sprinkle some icing sugar as garnish.

熱狗麵包(軟麵包) Soft Bread
形狀為橄欖型，質地相當柔軟。
通常中央會切開來擺放加熱過的
熱狗(又稱法蘭克福香腸)。

鮪魚鬆餅
Tuna fish Salad Waffle

主要材料：
鮪魚罐............................1罐

裝飾材料：
洋蔥.........................50公克
西芹.........................30公克
酸黃瓜......................30公克
巴西里........................5公克
鹽、白胡椒...................適量
美奶滋......................20毫升
蕃茄..............................1顆
蛋..................................1顆
黑橄欖...........................5顆

麵包：
鬆餅...............................1片

做法：
① 先將鬆餅煎好。
② 將主要材料及裝飾材料拌勻再放上鬆餅，並放上蕃茄、蛋片、黑橄欖裝飾即可。

Ingredients
Body：
Tuna fish canned 1Tin

Garnish：
Onion chopped 50gr
Celery chopped 30gr
Pickle chopped 30gr
Parsley chopped 5gr
S&P TO TASTE
Mayonnaise 80ml
Tomato diced 1EA
Hard Boiled Egg 1EA
Black olive 5EA

Bread：
Waffle 1p.c

Methods：
1. Cook waffle in waffle machine until golden brown.
2. Combine body and garnish items and mix well.
3. Place tuna fish salad on waffle. Garnish with black olive. Serve as need.

主要材料：

草莓	3顆
楊桃	1顆
哈蜜瓜	80公克
香蕉	80公克
奇異果	1顆
萊姆酒	20毫升
糖	40公克
檸檬皮碎	1顆

裝飾材料：

打發鮮奶油	30毫升

麵包：

鬆餅	1片

做法：

① 先將鬆餅煎好。

② 將主要材料拌勻，再放上鬆餅，並用鮮奶油及水果醬裝飾即可。

水果沙拉鬆餅
Fresh Fruit Salad on Waffle

Ingredients
Body：
Strawberry 3EA
Star Fruit 1EA
Honey dew melon 80gr
Banana 80gr
Kiwi 1EA
Rum 20ml
Sugar 40gr

Lemon zest chopped
1EA

Garnish：
Whipped Cream 30ml

Bread：
Waffle 1p.c

Methods：
1. Cook waffle in waffle machine until golden brown.
2. Combine all body items and mix well. Put fruits salad on waffle. Garnish with whipping cream and jam. Serve as need.

炒雞肉蘑菇鬆餅
Sautéed Chicken Meat and Mushroom Waffle

主要材料：
雞肉............................120公克

裝飾材料：
洋蔥塊........................50公克
洋菇............................30公克
香菇............................30公克
百里香........................1公克
玉桂葉........................1片
鮮奶油........................80毫升
鹽、白胡椒..................適量

麵包：
鬆餅............................1片

做法：
① 先將鬆餅煎好。
② 將主要材料及裝飾材料炒好放上鬆餅即可。

Ingredients
Body：
Chicken Meat 120gr

Garnish：
Onion diced 50gr
Mushroom Button 30gr
Mushroom shitake 30gr

Thyme 1gr
Bay leaf 1p.c
Cream 80ml
S&P TO TASTE

Bread：
Waffle 1p.c

Methods：
1. Cook waffle in waffle machine until golden brown.
2. Sauté all of the body and garnish items until cooked. Put cooked ingredients on waffle. Serve as need.

薄餅

Crêpes

Crêpes paste basic
薄餅麵糊　基本做法

材料：

麵粉	200公克	蛋黃	2顆
牛奶	350公克	鹽	2公克
蛋	5顆	融化奶油	20毫升

Ingredients

Flour 200gr
Milk 350gr
Egg 5EA

Egg yolk 2EA
Salt 2gr
Melted butter 20ml

做法：

① 麵粉過篩和鹽一起放入鍋盆，將蛋及蛋黃加入拌勻。

② 倒入牛奶，再加入融化奶油。

③ 將材料混合均勻成麵糊。

④ 將鍋子擦乾淨後加熱，取一勺麵糊倒入鋪成薄圓片狀。

⑤ 加熱至金黃色即可取出。

Methods :

1. Sieve flour and salt, mix well with egg and yolks.
2. Mix with milk.Combine with melted butter.
3. Mix well all the ingredients.
4. Clean the pan and heat.Take one ladle of batter and pour in pan.
   Make it round and thin.
5. Cook until golden brown.

78

奶油蘑菇雞肉袋
Chicken and Mushroom Cream in Crêpes bag

主要材料：
雞肉........................120公克
洋菇........................80公克

裝飾材料：
洋蔥........................50公克
蒜苗........................30公克
百里香......................1公克
玉桂葉......................1片
白葡萄酒...................50毫升
鮮奶油.....................80毫升
鹽、白胡椒.................適量

麵包：
薄餅........................6片

醬汁：
濃縮義大利甜醋........10毫升

Ingredients
Body：
Chicken diced 120gr
Mushroom Button diced
80gr

Garnish：
Onion chopped 50gr
Leek chopped 30gr
Thyme 1gr
Bay leaf 1p.c
White Wine 50ml
Cream 80ml
S&P TO TASTE

Bread：
Crepês 6p.c

Sauce：
Balsamic Reduced 10ml

做法：
① 先將薄餅煎好。
② 將主要材料及裝飾材料炒好，再用薄餅包起來成石榴球狀，再劃上濃縮義大利甜醋即可。

Methods：
1. Cook crêpes in pan until golden brown.
2. Sauté body and garnish items until cooked. Wrap all the cooked ingredients with crêpes. Make it into a bag shape. Garnish with balsamic reduction. Serve as need.

義式甜醋 Balsamic Vinegar
是義大利出產的陳年葡萄酒醋，依照義大利當地法令必須陳釀至少12年，等級更高的醋甚至有25年以上的佳釀。醋液本身顏色為黑色且相當濃稠，表面帶有些許釉亮的光澤，口味酸帶甜，相當圓潤。

79

炒牛肉薄餅
Sautéed Beef and Mushroom on Crêpes

主要材料：
牛肉切片 100公克

裝飾材料：
洋蔥塊.....................50公克
洋菇 30公克
香菇 30公克
鮑魚菇......................30公克
白葡萄酒 50毫升
玉桂葉 1片
鮮奶油......................80毫升
鹽、白胡椒.................適量

麵包：
薄餅 2片

做法：
① 先將薄餅煎好備用。
② 再將主要材料及裝飾材料炒熟，於盤中先放一片薄餅再放上炒好的牛肉，再蓋上一片薄餅即可。

Ingredients
Body：
Beef Tenderloin sliced
100gr

Garnish：
Onion chopped 50gr
Mushroom Button 30gr
Mushroom shitake 30gr
Mushroom Abalone
30gr

White Wine 50ml
Bay leaf 1p.c
Cream 80ml
S&P TO TASTE

Bread：
Crêpes 2p.c

Methods：
1. Cook crêpes in pan until golden brown.
2. Sauté all of the body and garnish items until cooked. Place one crêpes in the plate, then put some cooked ingredients, place another crêpes to cover. Serve as need.

水果沙拉薄餅

Fresh Fruit in Crêpes

主要材料：

草莓 3顆
楊桃 1顆
哈蜜瓜 80公克
香蕉 80公克
奇異果 1顆
萊姆酒 20毫升
糖 40公克
檸檬皮碎 1顆

裝飾材料：

打發鮮奶油...80毫升

麵包：

薄餅 6片

做法：

① 先將薄餅煎好。

② 將主要材料拌勻，加入打發鮮奶油，再捲入薄餅即可。

Ingredients
Body：
Strawberry 3EA
Star Fruit 1EA
Honey dew melon 80gr
Banana 80gr
Kiwi 1EA
Rum 20ml
Sugar 40gr
Lemon zest chopped 1EA

Garnish：
Whipped Cream 80ml

Bread：
Crêpes 6p.c

Methods：

1. Cook crêpes in pan until golden brown.
2. Combine all the body items and mix well. Add whipping cream and roll with Crêpes. Serve as need.

可麗餅

Crispy Crêpe

可麗餅麵糊　基本做法

材料：

高筋麵粉	100公克	蛋	1個
低筋麵粉	100公克	香草精	5ml
糖粉	40公克	水	120毫升
鹽	1公克	牛奶	200毫升
泡打粉	5公克	沙拉油	20公克

Ingredients

Bread flour 100gr
Cake flour 100gr
Icing sugar 40gr
B.P. 5gr
Salt 1gr

Egg 1EA
Salad oil 20gr
Vanilla extracts 5ml
Water 120gr
Milk 200ml

做法：

① 麵粉過篩和鹽、糖、泡打粉一起放入鋼盆，將蛋加入拌勻。
② 倒入水、牛奶混合。
③ 再加入香草精及沙拉油。
④ 將材料混合均勻成麵糊。
⑤ 將鍋子擦乾淨後加熱。
⑥ 取一勺麵糊倒入。
⑦ 用木棍推均成薄薄一層。
⑧ 加熱至金黃色即可翻起對折。
⑨ 再對折成三角狀即可。

Methods :

1. Sieve flour and combine with salt、sugar and BP, mix well with egg.
2. Pour water and milk, mix well.
3. Add vanilla extract and salad oil.
4. Mix well all the ingredients. And make into batter.
5. Clean the pan and heat.
6. Take one ladle of batter and pour in pan. Use a wooden bar to make it round and thin.
7. Cook until golden brown.
8. Pick it up.
9. And fold into a triangle.

81

芒果冰淇淋可麗餅
Mango Ice Cream in Crispy Crêpes

主要材料：
芒果冰淇淋 2球

裝飾材料：
覆盆子醬 20毫升
打發鮮奶油............80毫升
甜桃片 1/3顆

麵包：
可麗餅 1片

做法：
① 先將可麗餅煎好。
② 將芒果冰淇淋擠好放入
　可麗餅袋即可。

Ingredients
Body：
Mango Ice Cream 1
scoops

Garnish：
Raspberry Sauce 20ml
Whipped Cream 80ml
Peach Chips 1/3EA

Bread：
Crispy Crêpes 1p.c

Methods：
1. Cook crêpes in pan until golden brown.
2. Put a scoop of mango ice cream into the crêpe bag. Serve as need.

鮪魚可麗餅
Tuna fish Salad in Crispy Crêpes

主要材料：
鮪魚罐.............................1罐

裝飾材料：
洋蔥.............................50公克
西芹.............................30公克
酸黃瓜.............................30公克
巴西里.............................5公克
鹽、白胡椒.............................適量

麵包：
可麗餅.............................1片

抹醬：
美奶滋.............................80毫升

做法：
① 先將可麗餅煎好。
② 將主要材料及裝飾材料加入美奶滋拌勻即可，捲入可麗餅即可。

Ingredients
Body：
Tuna fish canned 1Tin

Garnish：
Onion chopped 50gr
Celery chopped 30gr
Pickle chopped 30gr
Parsley chopped 5gr
S&P TO TASTE

Bread：
Crispy Crepês 1p.c

Spread：
Mayonnaise 80ml

Methods：
1. Cook crêpes in pan until golden brown.
2. Combine body and garnish items with mayonnaise. Mix well. Roll with crêpe. Serve as need.

巧克力可麗餅

Chocolate mousse with Ice Cream in Crispy Crêpes

主要材料：
巧克力慕斯 ..50毫升

裝飾材料：
冰淇淋 2球

麵包：
可麗餅 1片

做法：
先將可麗餅煎好，再抹上巧克力醬，再放入冰淇淋，捲起來即可。

Ingredients
Body：
Chocolate mousse 50ml

Garnish：
Ice Cream (your choice) 2Scoop

Bread：
Crispy Crepês 1p.c

Methods：
Cook crêpes in pan until golden brown. Spread some chocolate syrup on crêpe. Roll it up and serve.

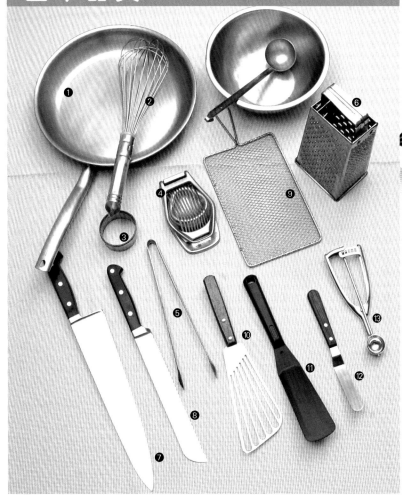

熱三明治機 Sandwich Griller 在將夾了餡的土司放入機器後，蓋起稍待幾分鐘後，土司便會被壓製成兩個小的三角狀的夾餡麵包，內部的加熱系統也會將土司表面加熱上色。

鬆餅機 Waffle Iron 拜科技進步之賜，在市面的3C賣場都可以買到這種操作容易且便宜的鬆餅機，除了這些家用機種外，大型餐飲機構中也可見比較專業且爐口多的機種，可在短時間內做出大量的產品。

❶ 斜邊翻炒鍋 Slope-Side Sauté Pan
這只西式的翻炒鍋具有傾斜的邊緣，可供廚師在加熱食物的同時，藉由翻炒的動作來使食物均勻與醬汁或調味料混合。此鍋也可用來煎(Pan fry)大塊的食物。

❷ 打蛋器 Wire Whips
藉由末端網狀的結構來使蛋液或是其他材料均勻混合，也可用來拌打鮮奶油或是麵糊。

❸ 圓切割器 Round Cutter
以鐵片圍成的圓模型，有不同的大小，可用來壓切較軟的食材，例如麵包或是蔬菜類。市面上除了圓形，也可買到方形、心形等不同形狀的切割器。

❹ 切蛋器 Egg Slicer
此器具利用彈簧和細鐵絲的結合，來切割軟而細緻的水煮蛋，本身設計為可橫切也可縱切。

❺ 夾子 Tong
在廚房中廣泛使用的器具，主要用來夾取食材。有些尖端比較小的夾子可用來擺放比較小的裝飾物(Garnish)。

❻ 研磨器 Box Grater
長方桶狀的研磨器具有四種不同的面，可將食材磨出不同大小、形狀和粗細的片、末或是絲。常用來處理乳酪、蔬菜等食材。使用後要仔細清洗，否則食物殘渣容易卡在細縫中滋生細菌害蟲。

❼ 主廚刀 Chef's Knife
是西餐廚房中通用性最高、用途最廣、尺寸最多的刀具，尖且細的刀尖可用來切割細緻的食材或是刻畫花紋；直且帶有弧度的刀身可用來切片切絲切丁，使用度極高；刀根的部分可用來剁切較硬的食材。

❽ 麵包刀：鋸齒刀 Serrated Knife
刀身具有鋸齒狀刀刃，可以拉切的方式切割柔軟且有組織的食材，例如蛋糕或是麵包等。

❾ 細篩網 Fine Sieve
利用鐵絲交叉所產生的細孔，可濾除麵粉或糖粉中的塊狀物或是不良雜質。本書中是用來將水煮蛋過篩，以取得細緻且粗細均勻的蛋末。

❿ 煎鏟 Offset Spatula
用來將加熱中的食物翻面或是鏟起煎好的食材，寬且平的前端可以用來鏟起細緻且易碎的食材。

⓫ 耐熱塑膠煎鏟
Flame Proof Plastic Spatula
材質為耐熱塑膠，用來將加熱中的食物翻面或是鏟起煎好的食材，寬且平的前端可以用來鏟起細緻且易碎的食材。

⓬ 三明治抹刀 Sandwich Spreader
和平抹刀一樣，可用來將抹醬或是材料抹在麵包上。稍具彎度的刀身，可適應各種角度。

⓭ 冰淇淋杓 Ice Scoop
有不同的大小尺寸，內部半圓形的特殊結構可以使冰淇淋順利脫離杓子。市面上也有無特殊結構的杓子。使用時，可將杓子泡在水中來提高溫度，以利挖取冰淇淋。

Easy Cook

三明治、鬆餅&醬料100道

作者　陳寬定

出版者 / 大境文化事業有限公司　T.K. Publishing Co.

發行人　趙天德　　總編輯　車東蔚

文案編輯　編輯部

美術編輯　R.C. Work Shop　　攝影　TOKU　CHAO

台北市雨聲街77號1樓

TEL：(02)2838-7996　　FAX：(02)2836-0028

法律顧問　劉陽明律師　名陽法律事務所

初版日期　2006年4月

定價　新台幣260元

ISBN 957-0410-53-1　　書　號　E60

讀者專線　(02)2836-0069

www.ecook.com.tw

E-mail　service@ecook.com.tw

劃撥帳號　19260956 大境文化事業有限公司

感謝以下廠商的協力配合，讓本書更豐富精彩，特此致謝！

十代食品行　07-3800678

麥香魚海產股份有限公司　07-8153216

勝源青果行　07-3226927

美國肉類出口協會　02-27361200

三明治、鬆餅&醬料100道

陳寬定 著：--初版.--臺北市：大境文化，

2006[民95]面；　公分. ----(Easy Cook：60)

ISBN 957-0410-53-1

1. 食譜 - 速食

427.14　　　　95004122

義式濃縮咖啡大全Espresso Book

日本Espresso咖啡冠軍

收錄了本書作者---門脇洋之，也是日本咖啡師大賽冠軍、世界咖啡師大賽第7名，被譽為日本Espresso咖啡達人多年來關於義式濃縮咖啡所有的知識與經驗。分為三個部分：1.義式濃縮咖啡(Espresso)的基本技術：研磨咖啡豆/填壓/咖啡機的設定/抽出Espresso/咖啡機的清理/打蒸氣奶泡/咖啡豆混合與烘焙知識。2.基礎咖啡&花式咖啡：濃縮咖啡及拉花52種/詳細配方及步驟圖解。3.從無到有開店實錄CAFÉ ROSSO：修習蛋糕製作/義大利朝聖/創業企劃書/銀行的融資店舖設計…等。對於喜愛義式濃縮咖啡的同好，能有全面而深入的瞭解，開業經營咖啡館的朋友，更可從中得到目前最實用、也最專業的Espresso終極品質技巧。

出版：大境文化
作者：門脇洋之
尺寸：15 × 21cm192 頁
定價：NT$340

經典調酒大全My standard cocktail

日本3位頂尖調酒達人的傾囊相授

以介紹標準雞尾酒為主更加入了原創雞尾酒，由3位調酒師(Bartender)來選酒，並將重點放在如何靈活而有技巧地發揮基酒的特性，及調製時的訣竅…等實際運用的層面上，同時加註3位調酒師各自的觀點與解說。

不同的調酒師，即使是調製同一種雞尾酒，也會調製出完全迥異的風味來。如果想要調製出吸引人的雞尾酒，除了必須對酒具有豐富的相關知識及熟練的技術之外，還須具備敏銳的五感：視覺、聽覺、嗅覺、味覺、觸覺，及成功地表現各種雞尾酒的能力。無論您是要更加精進自身調製雞尾酒的技巧，或是為了更充分享受品味雞尾酒的樂趣，我們都衷心地期待本書能夠成為您最佳的幫手。

出版：大境文化
作者：田中利明/永岡正光/内田行洋
尺寸：15 × 21cm192 頁
定價：NT$340

沿 虛 線 剪 下 ✂

大境文化信用卡訂書單

傳真專線： (02) 2836-0028

請放大影印後傳真

持卡人姓名：

生日：　年　月　日

身份證字號：□□□□□□□□□□

性別：□男　□女

聯絡電話：(日)　　　　(夜)　　　　(手機)

e-mail：

訂　購　書　名	數量(本)	金額

訂書金額：NT$　　　＋郵資：NT$ 80 (2本以上可免) ＝NT$

總訂購金額：　　　仟　　　佰　　　拾　　　元整
(請用大寫)

通訊地址：□□□ - □□□

寄書地址：□□□ - □□□

發卡銀行：　　　　□ VISA　□ Master

信用卡反面 後3碼：　　　□ 聯合卡　□ JCB

信用卡號：□□□□ - □□□□ - □□□□ - □□□□

有效期限：　　月　　　年

授權碼：
(免填寫)

商店代號：
(免填寫)

持卡人簽名：
(與信用卡一致)

發票：□二聯式　□三聯式　發票抬頭：

統一編號：□□□□□□□□

填單日期：　年　月　日

另有劃撥帳號可購書 / 19260956 大境文化事業有限公司

我們將儘速以掛號寄書，進度查詢專線：(02) 2836-0069 趙小姐

沿虛線剪下

台北郵政　73-196　號信箱

大境(出版菊)文化　　收

姓名：　　　　　　電話：

地址：